Lecture Notes in Control and Information Sciences

Edited by M. Thoma and A. Wyner

For information about Vols. 1–61 please contact your bookseller or Springer-Verlag.

Lecture Notes in Control and Information Sciences

Edited by M. Thoma and A. Wyner

110

B. J. Daiuto, T. T. Hartley,
S. P. Chicatelli

The Hyperbolic Map
and Applications to the
Linear Quadratic Regulator

Springer Science+Business Media, LLC

Authors
Brian J. Daiuto
Tom T. Hartley
Stephen P. Chicatelli

Department of Electrical Engineering
The University of Akron
Akron, Ohio 44325
USA

ISBN 978-3-540-96741-5   ISBN 978-3-540-40900-7 (eBook)
DOI 10.1007/978-3-540-40900-7

2161/3020-543210   Printed on acid-free paper

To our parents

ABSTRACT

This research monograph gives a complete discussion of the theory of the discrete-time hyperbolic map. Both scalar and matrix representations are considered. The dynamics of the map are analyzed and discussions of stability, quasiperiodicity, and chaos are included. Several applications are discussed, the most important being the discrete-time linear time-invariant quadratic regulator. The results obtained from this analysis are then extended to the continuous-time linear regulator. A discussion of the linear quadratic regulator with negative state weighting provides some important insights into the general regulator theory. The results contained in this monograph should be accessible to the first-year graduate student or advanced senior undergraduate. Interested readers should also have a background in ODE's, difference equations, optimization theory, and/or digital control theory.

CONTENTS

CHAPTER ONE
INTRODUCTION

The fundamental purpose of this book is to describe the behavior of a dynamical system which, because it is nonlinear, renders most conventional analysis techniques useless. Undergraduate engineering students spend a majority of their time learning about linear systems, which are ruled by the genre of iteration map shown in Figure 1.1. Under the additional constraint of shift-invariance, such systems are conveniently described by methods based on the Laplace transform (continuous time) and the z-transform (discrete time).

If the mapping F is nonlinear, these transforms no longer apply and, as a result, the entire spectrum of analysis techniques based on them typically are of no use. We must essentially analyze the system starting at "ground-zero" using whatever clever insights that gain be obtained.

Chapters 2 and 3 of this book perform such an analysis for a one-dimensional form of the discrete-time Riccati equation, which is the hyperbolic mapping {1.1}.

$$y_{k+1} = \frac{ay_k + b}{cy_k + d} \qquad \{1.1\}$$

The variables a, b, c, and d assumed to be constants, this equation has an iteration map which looks like Figure 1.2. Following this analysis, Section 3.A presents an extension to the multidimensional case, in which case equation {1.1} becomes equation {1.2}.

$$Y_{k+1} = [\, AY_k + B \,][\, CY_k + D \,]^{-1} \qquad \{1.2\}$$

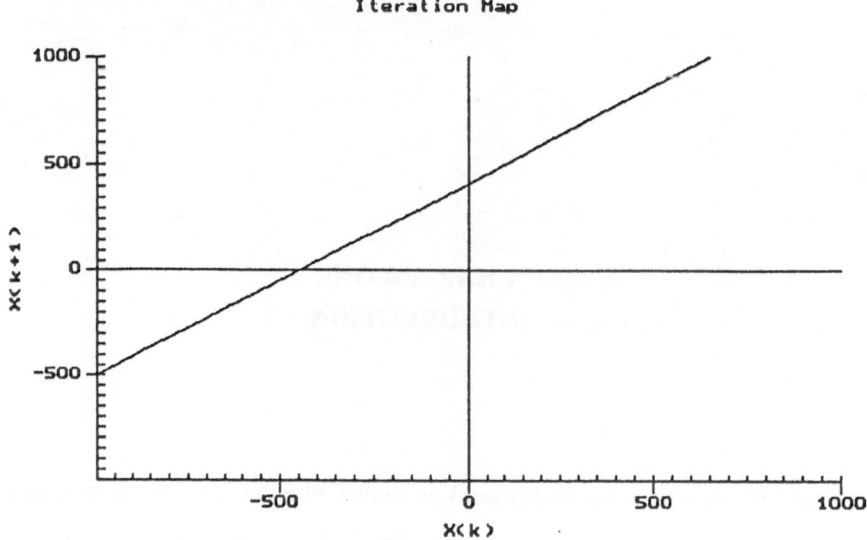

Figure 1.1

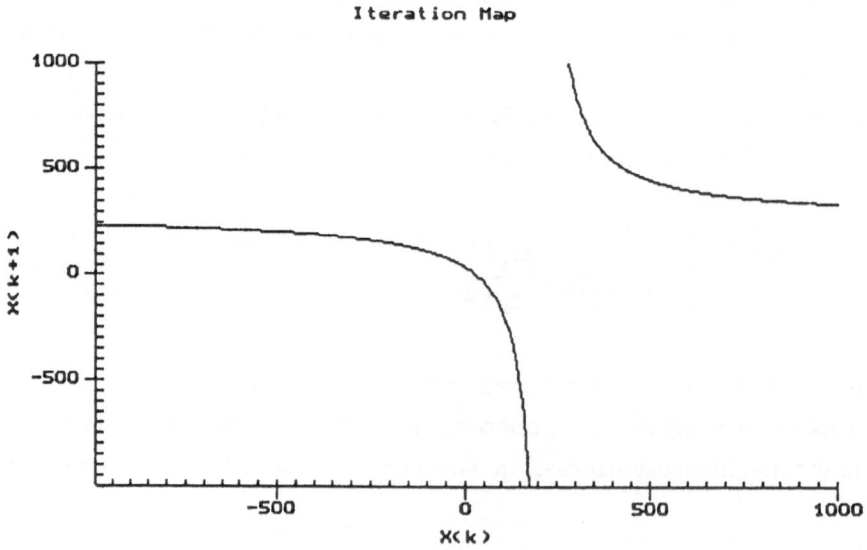

Figure 1.2

The variables A, B, C, and D are now all "n x n" matrices. Section 3.B then presents some practical applications of the difference equation {1.1}. The most useful application of {1.1} and {1.2}, however, is found in the analysis of the Linear Quadratic Regulator (LQR), which is introduced in Chapter 4.

Chapter 5 then uses the results of all previous chapters to investigate the LQR's behavior when one of its performance index parameters, "Q", takes on negative values. The principle conclusions of this investigation are (1) the resulting behavior of the system is quasiperiodic and *not* chaotic, (2) negative Q is probably not a very useful region of operation and (3), negative Q can probably not be used for evasion purposes.

Two comments regarding the syntax of this book are necessary. First, although they deal with the same general topic, Chapters 2 through 5 were designed to stand relatively independent of each other; conclusions regarding each respective topic are included in each chapter. Without too much loss of continuity, then, the reader may review the chapters in any order. Second, the subject of chaos arises in Chapters 2 and 5. Although not vital to the understanding of the main points of the book, the uninitiated reader is encouraged to first read the Appendix, which is a simplified introduction to the principle of chaos in physical systems.

CHAPTER TWO
QUALITATIVE DYNAMICS OF
THE HYPERBOLIC ITERATION MAP

In this section we examine the different trajectories of the hyperbolic iteration map, repeated here in equation {2.1}.

$$y_{k+1} = \frac{ay_k + b}{cy_k + d} \qquad \qquad \{2.1\}$$

Qualitative explanations of the behavior of this equation can be derived from studying the map of y(k+1) vs. y(k). In addition, quantitative expressions are derived for the necessary conditions for fixed-point and periodic trajectories.

We shall first examine fixed-point trajectories. The simplest case, as shown in Figure 2.1, is the straight decay into the fixed point. Figure 2.2 shows a similar path, except there is an initial "burst" before the trajectory falls into the fixed point. In addition, Figure 2.3 shows that the path can also oscillate around the fixed point as it falls into it. Finally, Figure 2.4 shows an initial burst followed by the damped oscillations around the fixed point.

These trajectories are easily explained in terms of the map of y(k+1) vs. y(k), given by equation {2.1}, a typical graph of which is shown in Figure 2.5. The graph is a hyperbola with vertical asymptotes at y(k) = -d/c, horizontal asymptotes at y(k+1) = a/c, and symmetry about a pair of lines oriented at 45 degrees and 135 degrees with respect to the positive x-axis (not necessarily passing through the origin).

If, in equation {2.1}, ad > bc, then the plot will lie in the second and fourth "quadrants" (considering the asymptotes as a new set of x' - y' axes). Figure 2.5 is one such example. If, on the other hand, ad < bc, the plot lies the in first and third quadrants, as in Figure 2.6. Finally, when ad = bc, equation {2.1} becomes

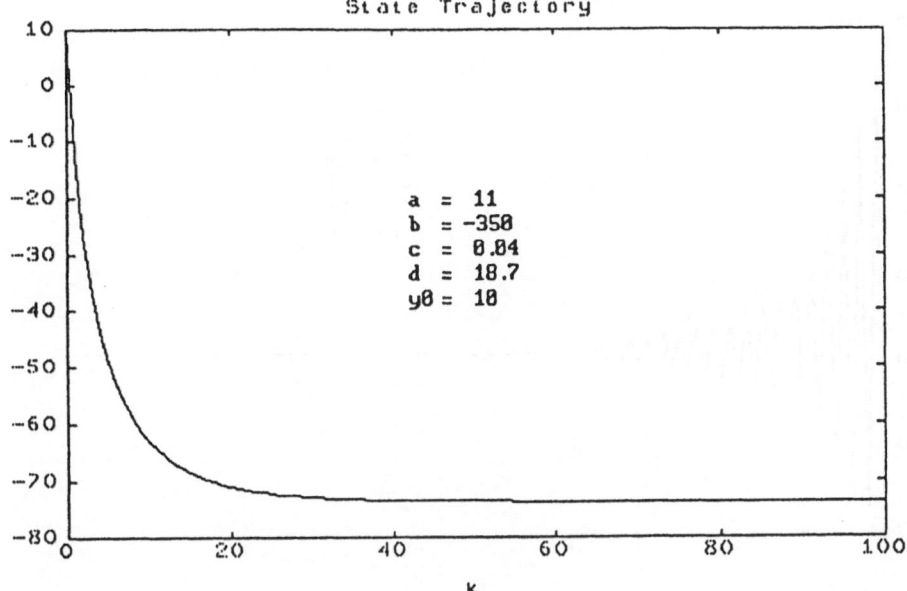

Figure 2.1

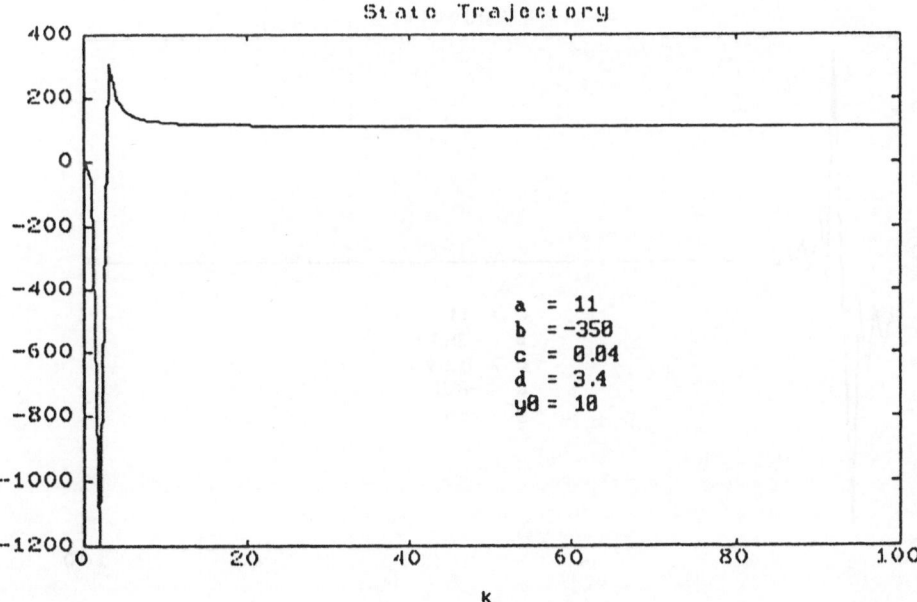

Figure 2.2

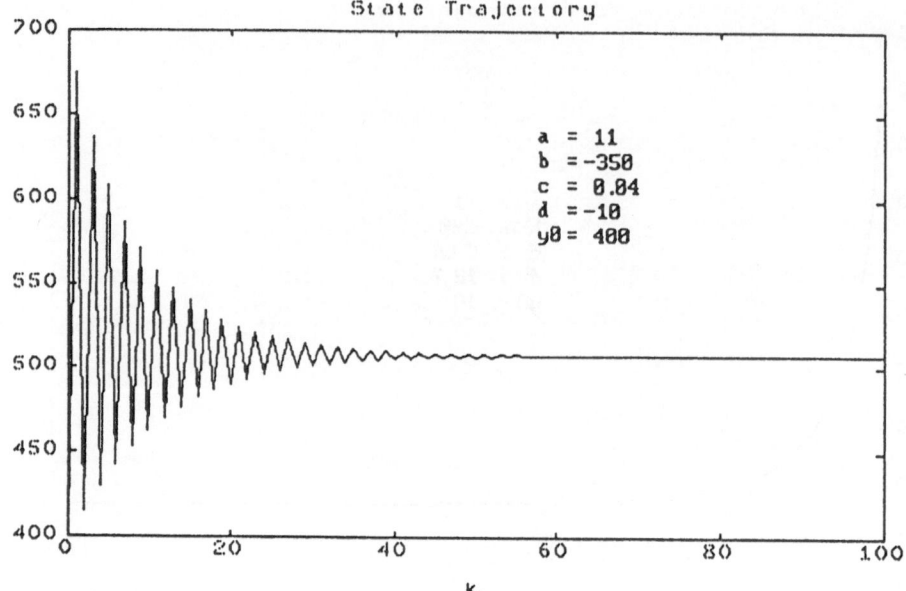

Figure 2.3

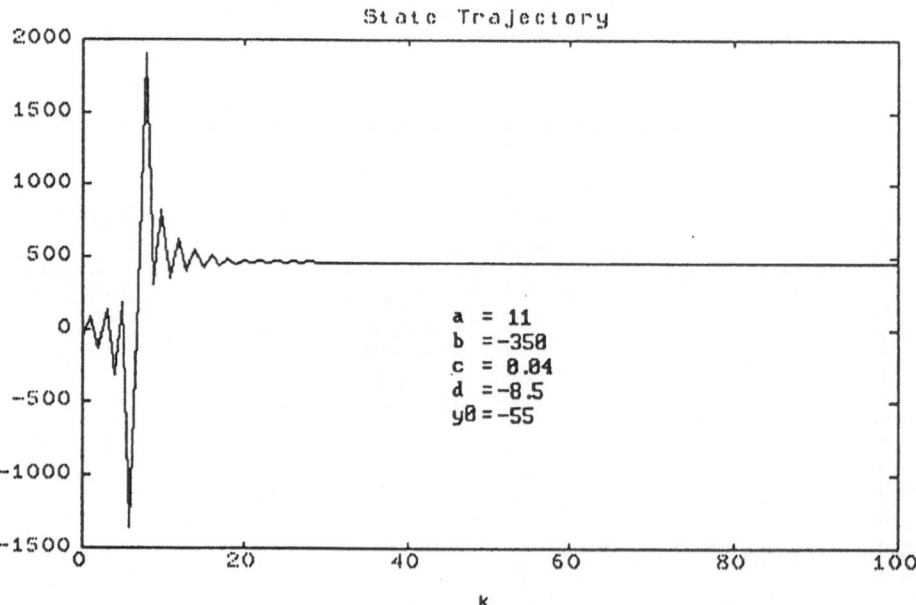

Figure 2.4

$$y_{k+1} = \frac{ay_k + b}{cy_k + d} = \frac{b(ady_k + bd)}{d(bcy_k + bd)} = \frac{b}{d} \qquad \{2.2\}$$

That is, the plot is simply a horizontal line at y(k+1) = b/d (with an undefined point at y(k) = -d/c).

In general, to examine trajectories of systems described by the map y(k+1) = f[y(k)], a vertical line is drawn from the initial point y(0) to y(1) = f[y(0)]. From this point, a horizontal line is drawn to the 45-degree line that passes through the origin; thus, the abscissa is now the point y(1). A vertical line is then drawn from this point to y(2) = f[y(1)], and so on. Figure 2.7 shows such a trajectory.

Such maps will be used to illustrate trajectories associated with fixed-point attractors as we derive them. For a fixed-point attractor, it is necessary that, for all k greater than some finite k',

$$y_{k+1} = y_k = y_{FIXED} \qquad \{2.3\}$$

In addition, it is necessary that the fixed point must "contract volume" around it, i.e. the Jacobian determinant of the mapping y(k) to y(k+1) must be less than unity. In our one-dimensional case, the Jacobian determinant is simply the magnitude of the derivative, thus, at the fixed point:

$$ABS\{ f'(y_{FIXED}) \} < 1 \qquad \{2.4\}$$

Solving {2.1} to fit condition {2.3} it is seen that

$$y_{FIXED} = \frac{ay_{FIXED} + b}{cy_{FIXED} + d} \qquad \{2.5\}$$

$$cy_{FIXED}^2 + (d-a)y_{FIXED} - b = 0 \qquad \{2.6\}$$

The solution to {2.6} is

$$y_{FIXED} = \frac{a-d}{2} \pm \frac{1}{2}\sqrt{[(a-d)^2 + 4bc]} \qquad \{2.7\}$$

Thus, when the descriminant of {2.7} is positive there will be two points which satisfy equation {2.5}. This can occure when:

8

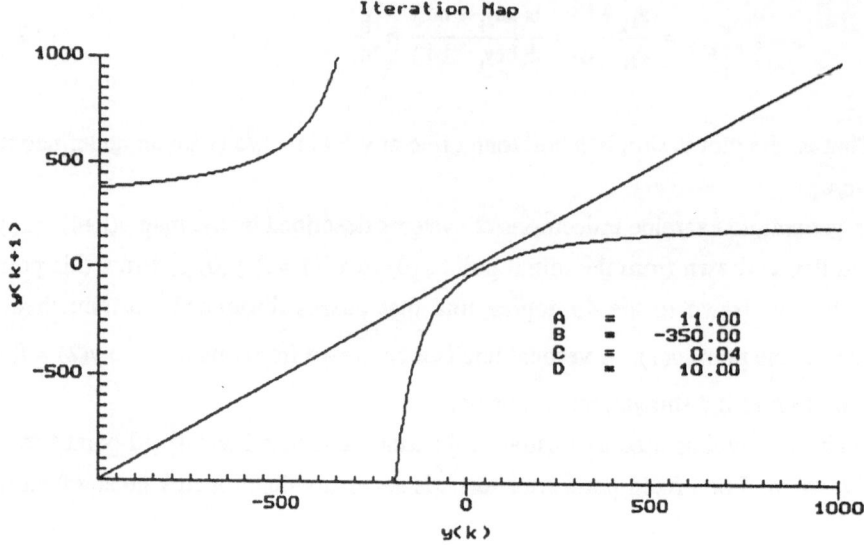

Figure 2.5

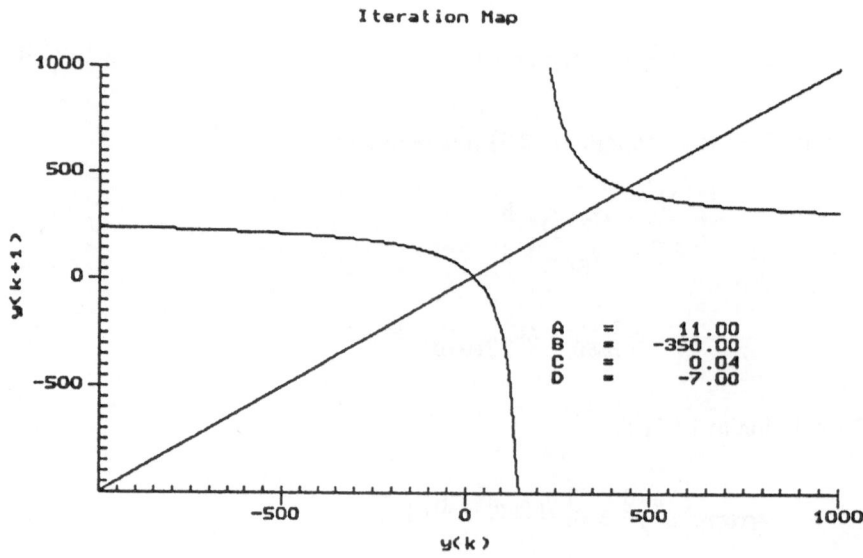

Figure 2.6

(1) bc > ad

(2) bc < ad

We examine each case separately.

(1)   For bc > ad, it is clear that the descriminant of {2.7} is always positive; a corresponding plot of y(k+1) vs. y(k) was shown in Figure 2.6. There are always two intersections of the map with the 45-degree line. Furthermore, it is seen from some mental shifting of the curve of Figure 2.6 that there will be exactly one attracting fixed point, i.e. one point will have slope magnitude greater than unity (it will repel) and the other will have slope magnitude less than unity (it will attract). Also note simply from the structure of Figure 2.6 that the basin of attraction of the single fixed point is the entire real axis (except the point y(0) = -d/c), since a point at large y(k) is "snapped back" very quickly to the center of the map.

Also for the case bc > ad we note that the slope at the fixed point is always negative, i.e. inbetween -1 and 0. Since the 45-degree line is inherently of positive slope, there exists an oscillatory convergence to the fixed point, as was depicted in Figure 2.7.

A unique feature or this attractor is its possible passage through a "burst" before settling into the fixed point. This occurs as a function of the initial condition. Figure 2.8 shows the trajectory of y(k) with the same parameters used as in Figure 2.7 except that a different initial condition is used. The "burst" is clearly seen to arise from the trajectory being "thrown over" to the other side of the vertical asymptote y = -d/c. The associated trajectories for Figures 2.7 and 2.8 are shown in Figures 2.9 and 2.10, respectively.

As a final note for the case bc > ad, it is found that, for a very limited set of parameters a, b, c, and d, there is a stable period-2 attractor wich is a function of initial conditions. The two examples in Figures 2.11 and 2.12 outline this. These figures show, for the same set of parameters a, b, c, and d, there is a different attractor of period 2. If the trajectory starts at a point y(0), it will stay on a trajectory which hits y(0) every other iteration. In short, every point y(0) is a period-2 attractor. This is easily explained by looking at the symmetry of Figures 2.11 and 2.12 around the 45-degree line. More rigorously, however, we take the second iterate of y(k) and set it equal to y(k).

$$f[\, f[\, y_k \,]\,] \;=\; y_k \qquad\qquad \{2.8\}$$

This yields the following necessary condition.

$$(a+d)\,[\, c y_k^2 \;+\; (d-a)y_k \;-\; b \,] \;=\; 0 \qquad\qquad \{2.9\}$$

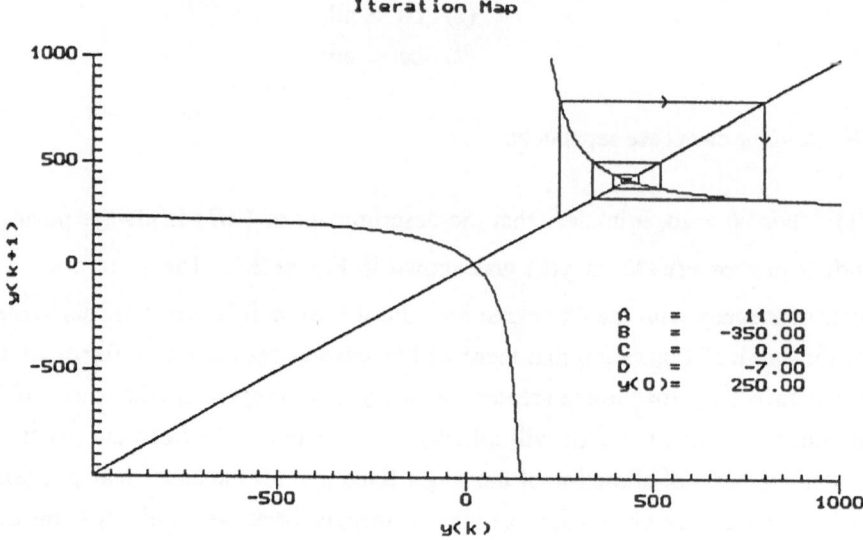

Figure 2.7

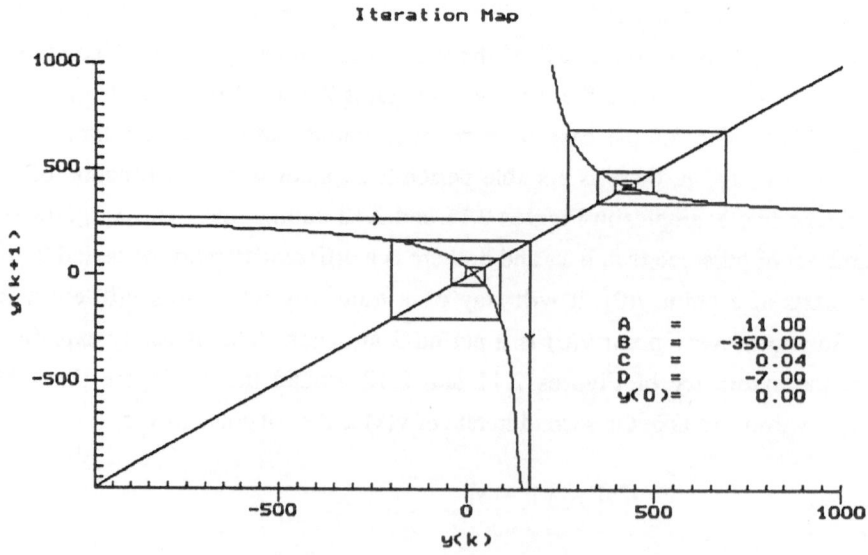

Figure 2.8

Noting that the bracketed quantity is simply the period-1 criterion of equation {2.6} (a period-1 cycle is also a period-2 cycle!) it is seen that the necessary condition is:

$$a + d = 0 \qquad \text{or} \qquad a = -d \qquad \qquad \{2.10\}$$

Since the vertical and horizontal asymptotes of the map {2.1} are at $y(k) = -d/c$ and $y(k+1) = a/c$, respectively, equation {2.10} simply verifies our intuition: there is a period-2 cycle, such that each point $y(0)$ lies on it, whenever there is a perfect symmetry of the map {2.1} about the 45-degree line (that is, when the asymptotes meet on the 45-degree line). In this case, the fixed points of {2.1} each have a Jacobian determinant of magnitude exactly unity, and hence they neither attract nor repel, but maintain the present orbit.

Due to the uniqueness of the symmetry involved, intuition dictates that this is the *only* periodic attractor for bc > ad (when the map is in the first and fourth "quadrants"). Clearly, if the parameters deviate from a = -d, the magnitude of the Jacobian determinant at one fixed point is driven below 1 and we have a period-1 again. Thus, for bc > ad, there is a fixed-point trajectory for all sets except when a = -d, when there is a period-2 attractor.

This indeed looks very strange on a density map versus the parameter a or d, whose form would have the structure of Figure 2.13. Note that the point d = -a the the parameter value for d which shifts the fixed point quadrant. Wrapping up this particular case, it is important to realize that the volume in a-b-c-d space occupied by the hyperplane a = -d is zero. The "probability" of observing such a period-2 is zero; even in digital computers, round-off errors causing a ≠ d can cause the trajectory to fall into a fixed point, although very slowly, as shown in Figure 2.14.

(2)　Turning to the case ad > bc, it is seen that the map {2.1} occupies the second and fourth "quadrants". There are two basic types of trajectories which occur in this case, the fixed-point and the oscillatory trajectory.

The fixed-point trajectory again occurs when equation {2.5} is satisfied, and corresponds to one of the two branches of the map {2.1} crossing the 45-degree line, as shown in Figure 2.15. Note by inspection that, as opposed to the case bc > ad, intersection with the 45-degree line does not necessarily have to occur. When intersection does occur, as dictated by the existence of real roots to {2.7}, there will arise one unstable and one stable point since, except for the case when the descriminant of {2.7} is zero, there must be two intersections with the 45-degree line. When the descriminant of {2.7} is zero, the situation of Figure 2.16 arises. The trajectory asymptotically approaches the fixed point from the "stable side", i.e. the side where the slope of the map is of magnitude less than unity. If the initial point is on the unstable side of the fixed point, it will first be "thrown over" to the

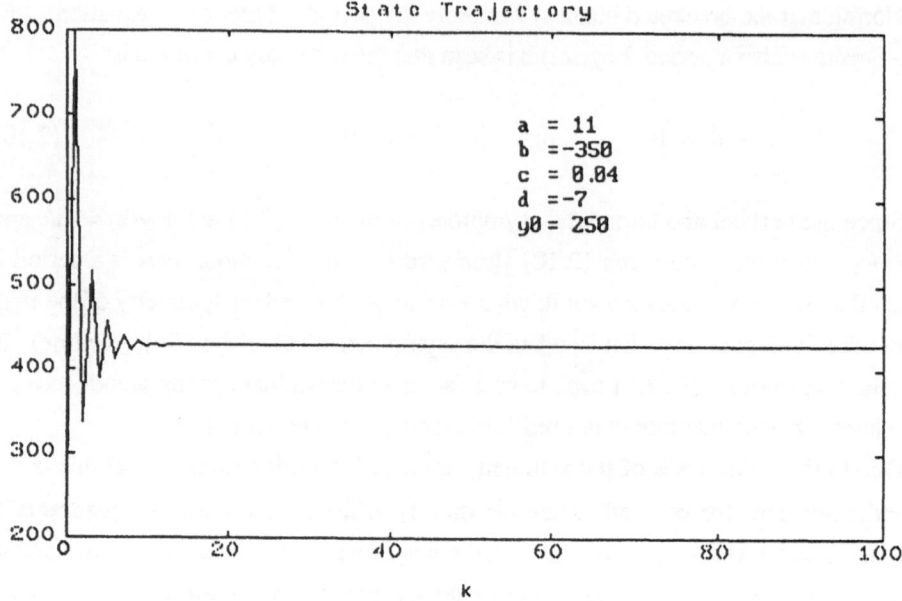

Figure 2.9

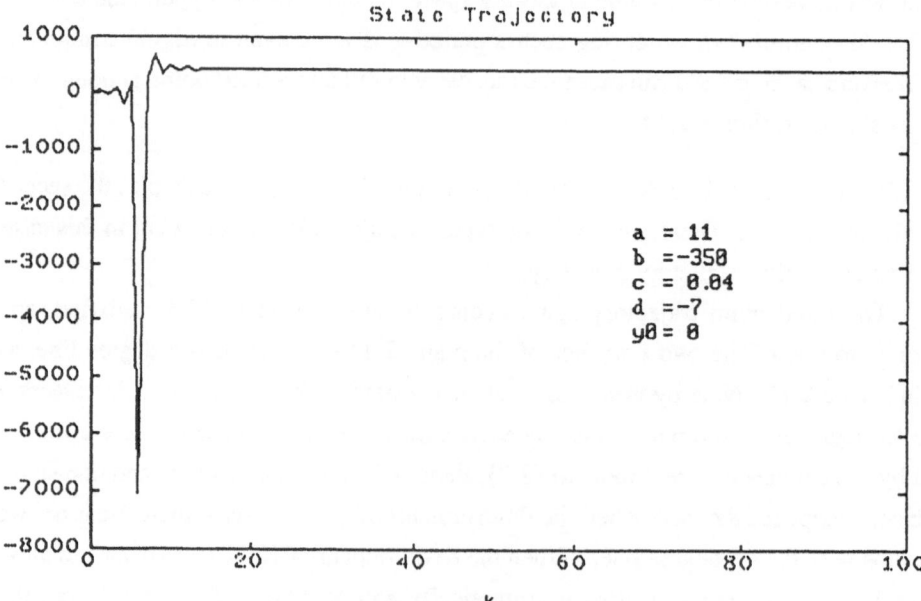

Figure 2.10

13

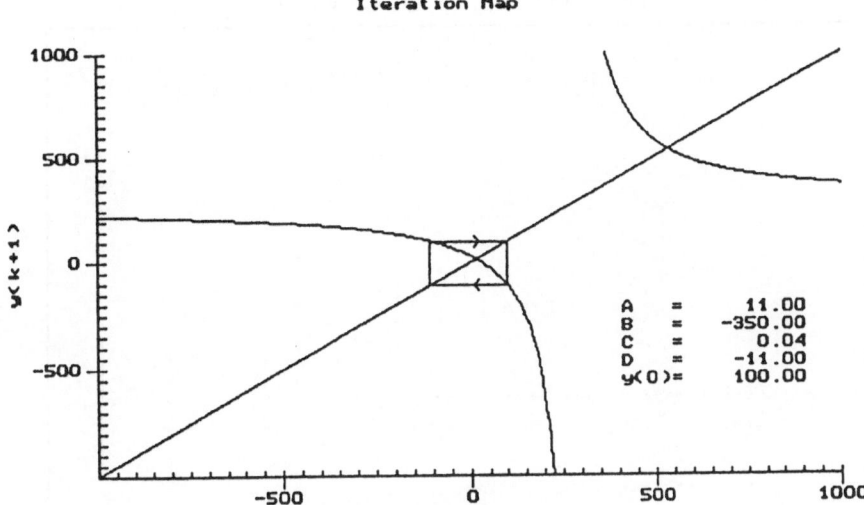

Figure 2.11

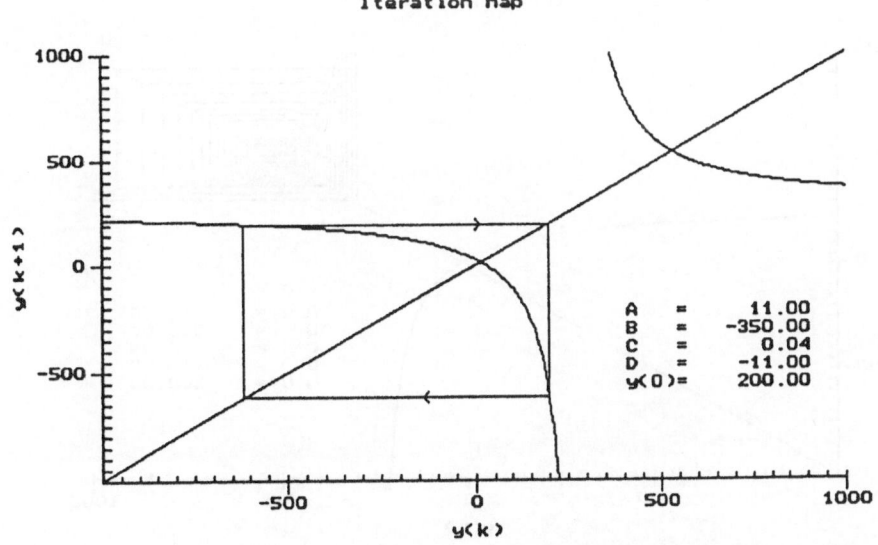

Figure 2.12

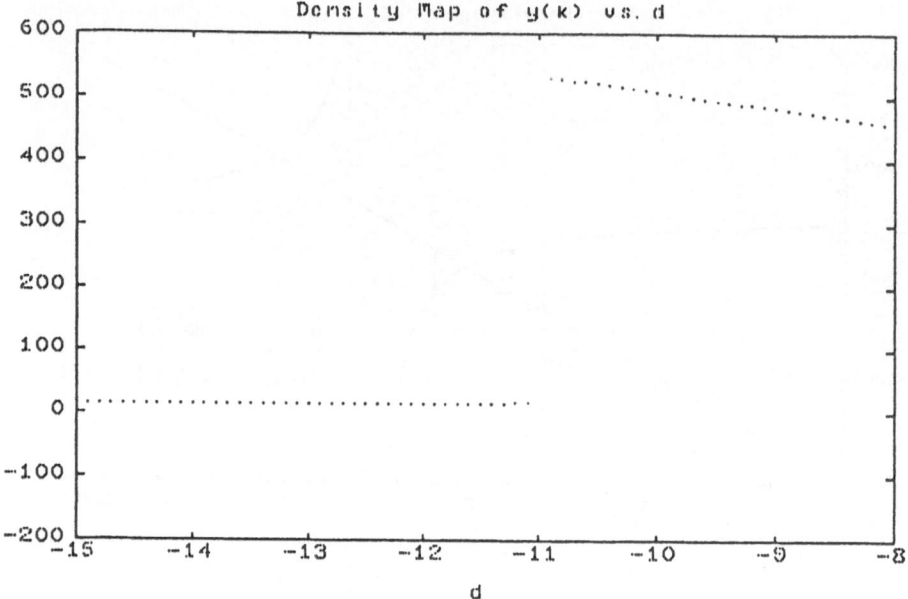

Figure 2.13

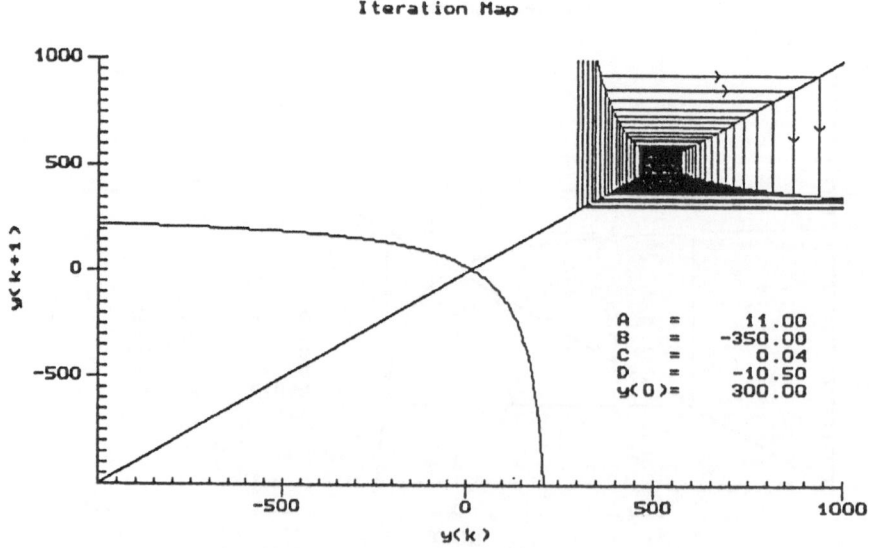

Figure 2.14

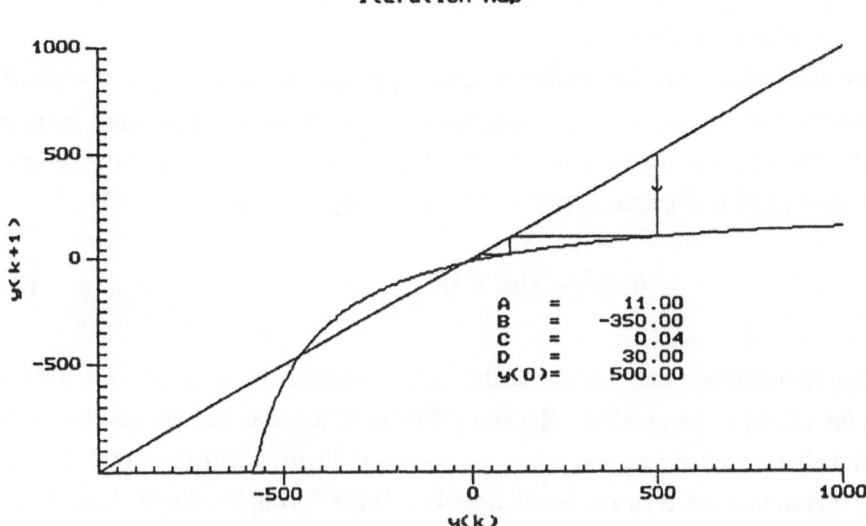

Figure 2.15

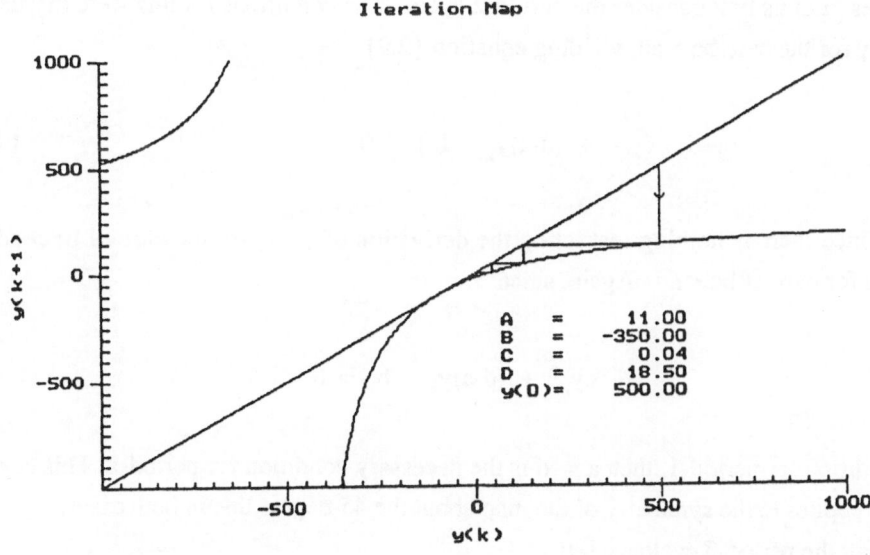

Figure 2.16

stable side by the asymptotic nature of the map. Corresponding trajectories for these cases were shown in Figures 2.1 and 2.2.

We shall now study the oscillatory case. Except for the very special case of the period-2 attractor previously discussed, oscillatory behavior occurs only when there is no intersection of the map {2.1} with the 45-degree line, or, equivalently, when the descriminant of {2.7} is negative. The case bc < ad is a subset of the case:

$$(a-d)^2 + 4bc < 0 \qquad \qquad \{2.11\}$$

This is significant only insofar as the map {2.1} must be in the second and fourth quadrants for {2.11} to be possible. Several different trajectories for the oscillatory case, along with their associated iteration maps, are shown in Figures 2.17 through 2.22. In the oscillatory region of a-b-c-d space, which does in this case occupy volume, the trajectories can only be classified in the rudimentary categories of "periodic" or "aperiodic", an example of the former being Figure 2.18 and examples of the latter being Figure 2.17 and 2.19. Other than looking at the fascinating figures generated by their iteration maps, there is little rigorous analysis possible for the aperiodic trajectories. Often called "snap-back repellers", their trajectories are characterized by bursts in magnitude which, when large enough, send the trajectory back into the center of the map.

On the other hand, some conclusions can be derived regarding the periodic trajectories. Let us first consider the period-2 cycle. The conditions for this were discussed previously for the case bc > ad, yielding equation {2.9}.

$$(a+d) [cy_k^2 + (d-a)y_k - b] = 0 \qquad \qquad \{2.9\}$$

Since there is nothing restricting the derivation of {2.9} to the case of bc > ad, it also holds for case of bc < ad. Again, since

$$cy_k^2 + (d-a)y_k - b = 0$$

is the condition for period-1, then a = -d is the necessary condition for period-2. This condition then applies to the symmetry of the map about the 45-degree line in both cases.

For the period-3 cycle we set:

$$f[f[f[y_k]]] = y_{k+3} = y_k \qquad \qquad \{2.12\}$$

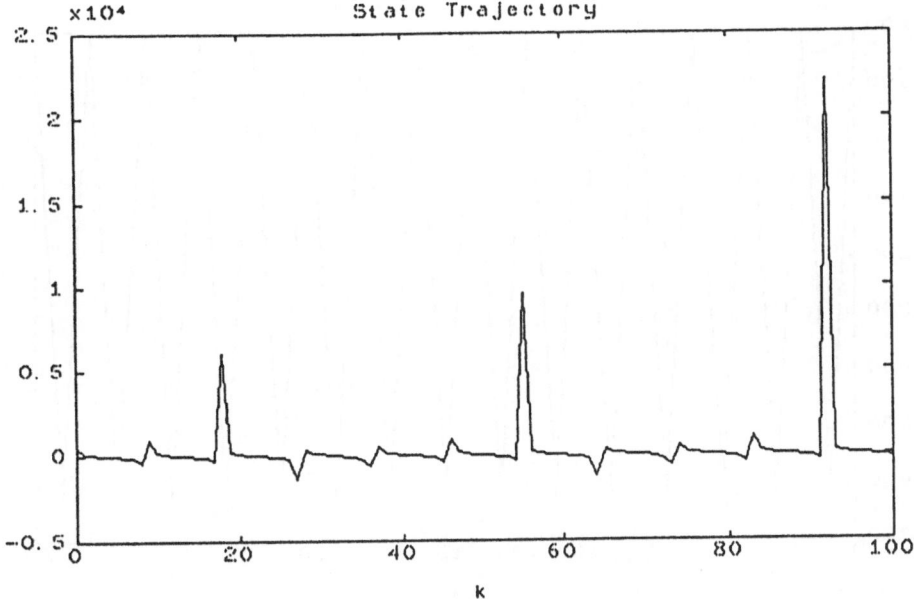

Figure 2.17

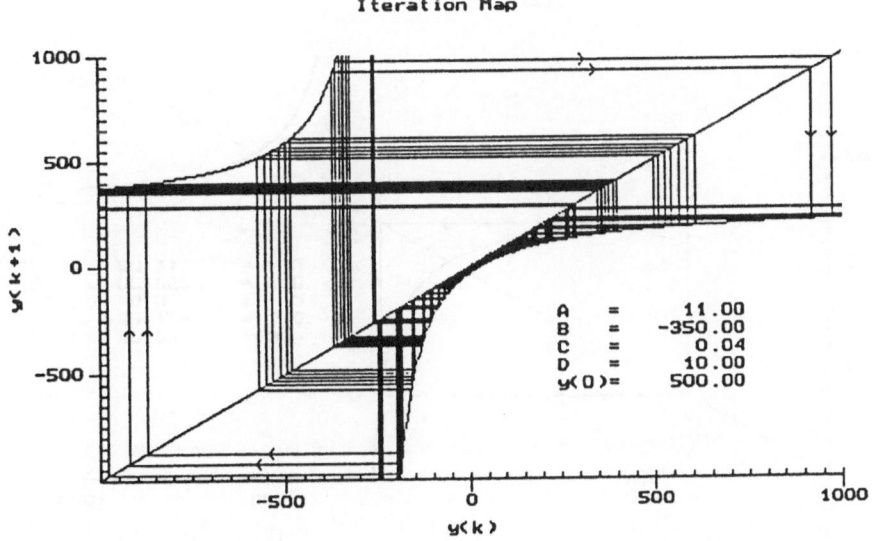

Figure 2.18

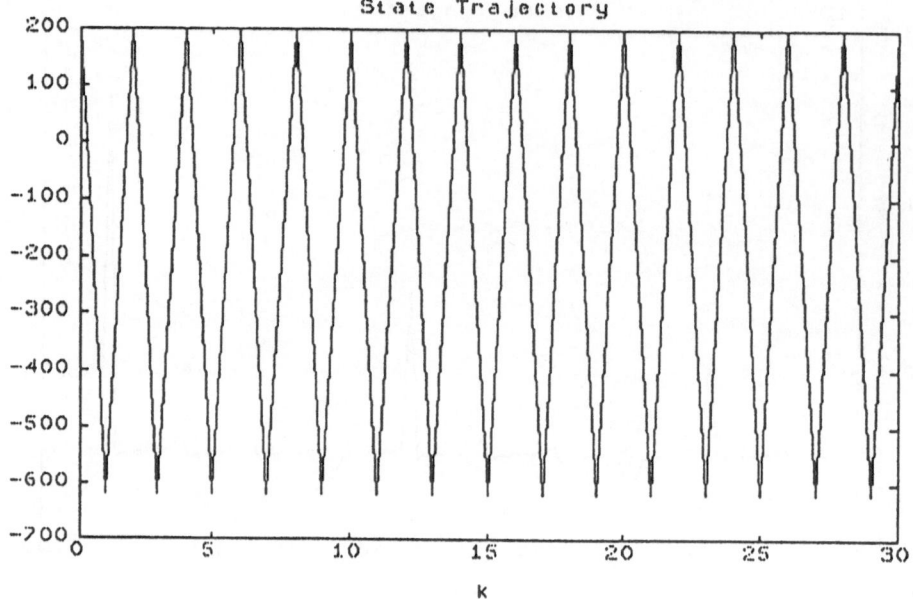

Figure 2.19

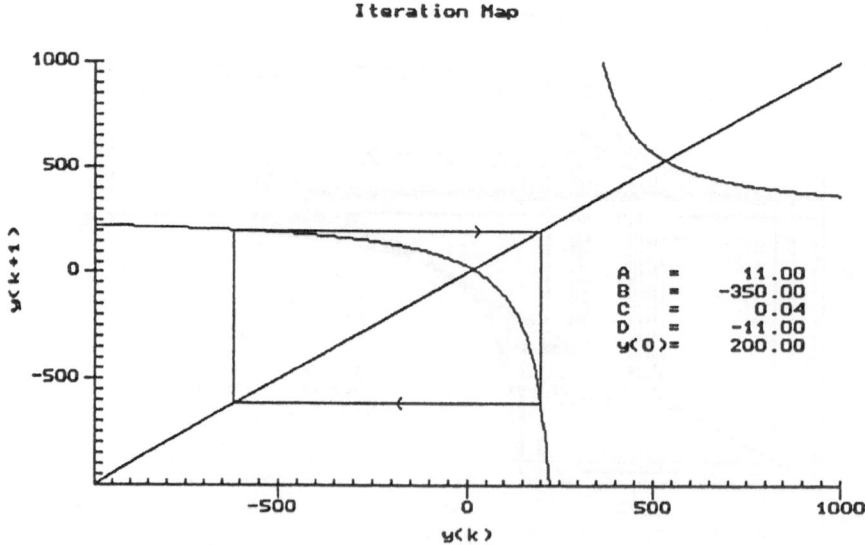

Figure 2.20

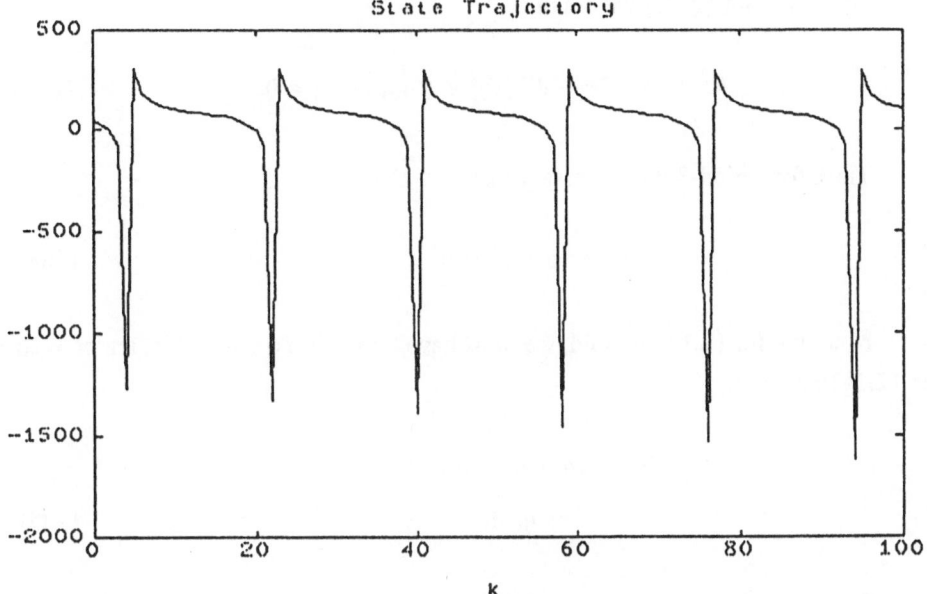

Figure 2.21

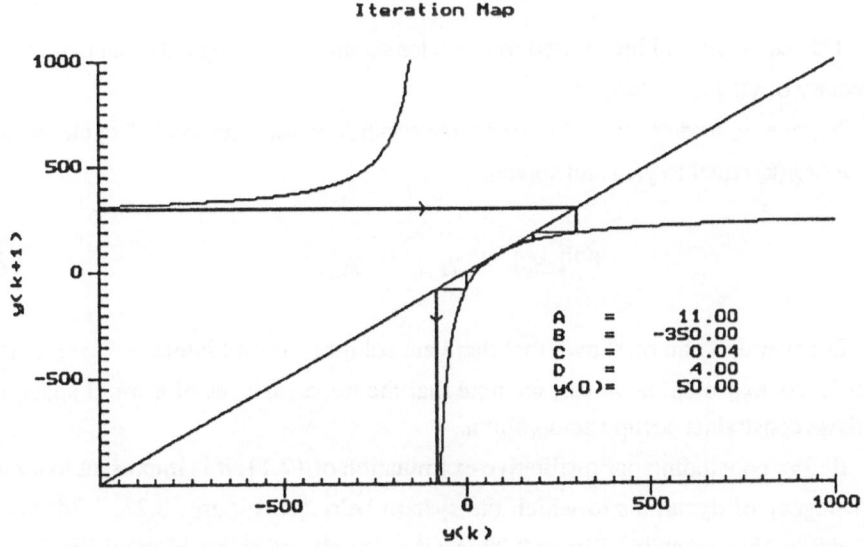

Figure 2.22

Upon solving {2.12} we obtain:

$$(a^2 + ad + bc + d^2)[cy_k^2 + (d-a)y_k - b] = 0 \qquad \{2.13\}$$

Thus, for period-3 it is necessary that:

$$a^2 + ad + bc + d^2 = 0 \qquad \{2.14\}$$

Note that for {2.14} to hold, we must have bc < ad. To prove this, let bc = ad + ∂. Then {2.14} becomes:

$$a^2 + 2ad + d^2 + \partial = 0$$
$$(a+d)^2 = -\partial \qquad \{2.15\}$$

Therefore, $\partial < 0$ and bc < ad. This shows that the map {2.1} must be in the second and fourth quadrants for a period-3 attractor to exist. Rearranging terms in {2.14}, it can also be shown that this is equivalent to {2.16}.

$$-\sqrt{3} = \frac{\sqrt{[-4bc - (a-d)^2]}}{a+d} \qquad \{2.16\}$$

This equation will be referred to in a later section, when an analytical expression for the trajectory of y(k) is developed.

In general, to determine the necessary conditions for a period "n" cycle we set the nth iterate of y(k) equal to y(k) and solve:

$$f^{(n)}[y_k] = y_{k+n} = y_k \qquad \{2.17\}$$

In general, it can be shown that there are solutions for all integers n greater than or equal to 1. As expected, however, we note that the hypersurfaces of a-b-c-d space which satisfy these constraints occupy zero volume.

Before concluding our qualitative examination of {2.1}, it is important to comment on the category of dynamics to which the system belongs. Figures 2.23, 2.24, and 2.25 help to address this question. Figure 2.23 is a density plot of points hit by the trajectory of {2.1} versus the parameter "d" for the case bc < ad. Clearly, there is a rather violent collision of the fixed point into the "oscillatory" region when

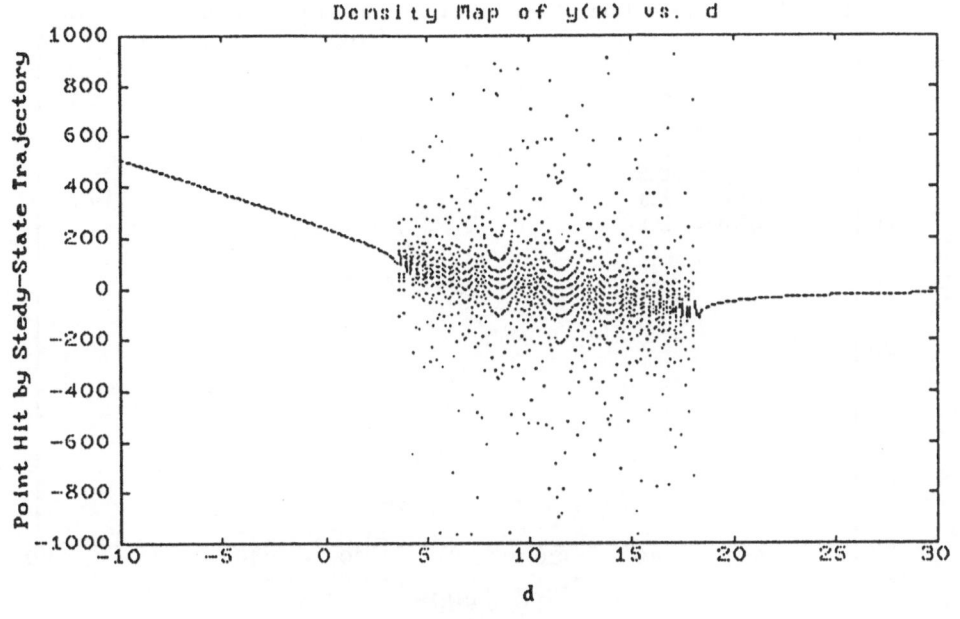

Figure 2.23

$$cy_k^2 + (d-a)y_k - b = 0.$$

Inside the region $(a-d)^2 + 4bc < 0$ the behavior is indeed very strange; all but a countably infinite (zero volume) set of values for "d" yield aperiodic trajectories. Yet there are values of "d" which yield trajectories of *any* integer period n, n≥1.

Athough this is indeed very peculiar behavior, is the system really "chaotic"? The answer is *no*, and the reason for this lies in the system's lack of sensitive dependence on initial conditions, which is the key characteristic of chaotic systems. This point is illustrated in Figures 2.24 and 2.25. To interpret Figure 2.24, consider a system of the form of {2.1} whose parameters a,b,c, and d are such that we are in the "strange" region of Figure 2.23. Now find the trajectory of the system for some initial condition y(0), denoted by y(k). Then choose another initial condition *very close* to y(0), denoted by y'(0), and find the associated trajectory, denoted by y'(k). Then proceed to plot y'(k) versus y(k) for all k; Figure 2.24 is the result.

Note that there is a definite relation between the two trajectories; clearly, an analytical expression for y'(k) can be developed in terms of y(k) if we are given y(0) and

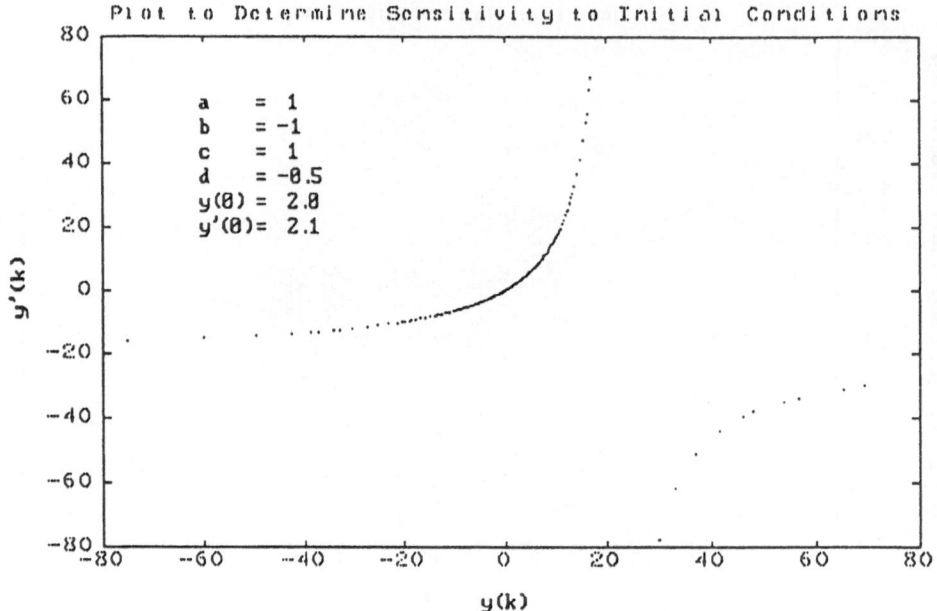

Figure 2.24

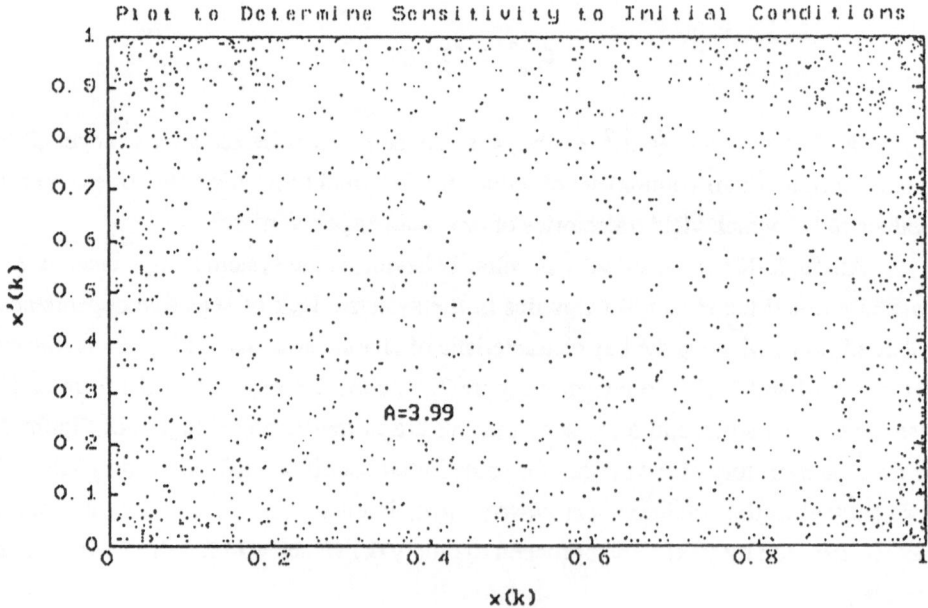

Figure 2.25

y'(0). If this were a truly chaotic system, this would be impossible. A physical example would be the ability to analytically relate the paths followed by two initially adjacent water particles as they travel down a river. If {2.1} for some reason modeled this situation then, although we would encounter some strange curlyques as we go down our river, we would never encounter turbulent rapids or waterfalls.

For a system to be truly chaotic, there must be a sensitive dependence on initial conditions: two points which were initially very close to each other will, under iteration, separate and, after a short time, follow such vastly different trajectories that, we are unable to determine that they even came from the same region. There should be no correlation between y'(k) and y(k) after a few iterations, and a plot of y'(k) vs. y(k) should yield an untenable blizzard of points. This is clearly the case for Figure 2.25, which was built from the well-known Logistic equation in its chaotic region. Figure 2.25 is a similarly constructed plot for the truly chaotic system {2.18}.

$$x_{k+1} = Ax_k(1 - x_k) \qquad A = 3.99 \qquad \{2.18\}$$

Clearly, there is no correlation between x'(k) and x(k).

In conclusion, then, our system {2.1} is deterministic. It is not chaotic. Nevertheless, the system is a nonlinear one which, as we have seen, exhibits a wide variety of peculiar behavior. Although the term "quasiperiodic" has varying uses, we shall refer to the system

$$y_{k+1} = \frac{ay_k + b}{cy_k + d} \qquad \{2.1\}$$

as being quasiperiodic, primarily because of the form of the density plot of Figure 2.23.

CHAPTER THREE
GENERAL SOLUTION DERIVATION FOR
THE HYPERBOLIC ITERATION MAP

In this section some principle results are derived; specifically, the trajectory of the equation

$$y_{k+1} = \frac{ay_k + b}{cy_k + d} \qquad \{3.1\}$$

for any initial condition y(0) not equal to -d/c for any set of parameters a, b, c, and d (c and d not both zero).

Clearly, the equation {3.1} is nonlinear, and its strange behaviors were discussed in Chapter 2. However, let us think of the number y(k) as a quotient defined by

$$y_k = \frac{u_k}{v_k} \qquad \{3.2\}$$

Equation {3.1} can then be rewritten

$$\frac{u_{k+1}}{v_{k+1}} = \frac{au_k/v_k + b}{cu_k/v_k + d} = \frac{au_k + bv_k}{cu_k + dv_k} \qquad \{3.3\}$$

It is apparent that if we solve for the trajectories of u(k) and v(k) then the trajectory of y(k) is known by {3.2}. Examining the form of {3.3} it is seen that we can rewrite it in a similar form {3.4}:

$$\begin{bmatrix} u_{k+1} \\ v_{k+1} \end{bmatrix} = \begin{bmatrix} a & b \\ c & d \end{bmatrix} \begin{bmatrix} u_k \\ v_k \end{bmatrix} \qquad \{3.4\}$$

From discrete-time linear system theory, the sequences $\{u(k)\}$ and $\{v(k)\}$ can be found in terms of the initial conditions $u(0)$ and $v(0)$ by using the z-transform method for a two-dimensional system. Specifically,

$$\begin{bmatrix} u_k \\ v_k \end{bmatrix} = \begin{bmatrix} a & b \\ c & d \end{bmatrix}^k \begin{bmatrix} u_0 \\ v_0 \end{bmatrix} = A^k \begin{bmatrix} u_0 \\ v_0 \end{bmatrix} \qquad \{3.5\}$$

where

$$A^k \text{ is also the inverse z-transform of } [\, z(zI - A)^{-1} \,] \qquad \{3.6\}$$

Letting

$$A^k = \begin{bmatrix} A11_k & A12_k \\ A21_k & A22_k \end{bmatrix}$$

and substituting into the above equations, we obtain

$$y_k = \frac{A11_k u_0 + A12_k v_0}{A21_k u_0 + A22_k v_0}$$

$$= \frac{A11_k \dfrac{u_0}{v_0} + A12_k}{A21_k \dfrac{u_0}{v_0} + A22_k}$$

$$y_k = \frac{A11_k y_0 + A11_k}{A21_k y_0 + A22_k} \qquad \{3.7\}$$

Clearly then, by viewing $y(k)$ as the quotient of two states of a regular linear system, we find that an analytical solution for its trajectory is possible. The system looks like Figure 3.1.

Before examining the specific solutions to $\{3.7\}$, it is important to note that, although it has been cast in a "linear" format, the system is indeed nonlinear and its stability cannot be determined from linear systems analysis techniques. In standard linear analysis, we would render the system of Figure 3.1 unstable if there were eigenvalues, i.e. solutions of

$$|zI - A| = 0 \qquad \{3.8\}$$

of magnitude greater than unity. However, since $y(k) = u(k)/v(k)$ then the value $y(k)$ can remain bounded even if both $u(k)$ and $v(k)$ increase without bound. In our solution of $\{3.7\}$, then, we are not concerned with the magnitudes of the roots of $\{3.8\}$. Rather, only the types (real, complex) of roots are important. There are three possible cases, each of which will be treated separately.

(1) Distinct real eigenvalues.
(2) Identical real eigenvalues.
(3) Complex eigenvalues.

(1) Distinct Real Eigenvalues

When, given the matrix

$$A = \begin{bmatrix} a & b \\ c & d \end{bmatrix} \qquad \{3.9\}$$

the values of a, b, c, and d are such that

$$(a-d)^2 + 4bc > 0 \qquad \{3.10\}$$

then the eigenvalues p1 and p2 of A are real and distinct, and are equal to

$$p1,p2 = \tfrac{1}{2}(a+d) \pm \sqrt{[(a-d)^2 + 4bc]} \qquad \{3.11\}$$

with p1 arbitrarily assigned to the value of the larger-magnitude root.
Using the inverse z-transform method to find A^k yields

$$A^k = \frac{1}{p1 - p2} \begin{bmatrix} (p1-d)p1^k - (p2-d)p2^k & b(p1^k - p2^k) \\ c(p1^k - p2^k) & (p1-d)p1^k - (p2-d)p2^k \end{bmatrix} \qquad \{3.12\}$$

and, using $\{3.7\}$, we find the trajectory is

$$y_k = \frac{y_0[(p1-d)p1^k - (p2-d)p2^k + b(p1^k - p2^k)]}{y_0 c(p1^k - p2^k) + (p1-a)p1^k + (p2-a)p2^k} \qquad \{3.13\}$$

Upon examination of {3.13} it is seen that the system will experience some "transients" at first if the denominator approaches zero for some iterations. After this, however, a steady-state value is reached whose analytical expression is given by {3.14}.

$$y_{ss} = \lim_{k \to \infty} \frac{y_0(p1 - d)(p1/p2)^k + b}{y_0c(p1/p2)^k + (p1 - a)(p1/p2)^k} \qquad \{3.14\}$$

$$= \frac{y_0(p1 - d)}{y_0c + p1 - a}$$

Further simplification and use of {3.11} yields

$$y_{ss} = \frac{p1 - d}{c} \qquad \{3.15\}$$

Some typical trajectories for the system in this "overdamped" case (eigenvalues of A real and distinct) are shown in Figures 3.2 through 3.4.

(2) Repeated Real Eigenvalues

When the values a, b, c, and d are such that

$$(a - d)^2 = -4bc \qquad \{3.16\}$$

then the eigenvalues of A are repeated and real, equalling

$$p = p1 = p2 = 0.5(a + d) \qquad \{3.17\}$$

Again using the z-transform method, A^k is given by equation {3.18}.

$$A^k = p^k \begin{bmatrix} 1 + k\frac{a-d}{a+d} & k\frac{2b}{a+d} \\ k\frac{2c}{a+d} & 1 + k\frac{d-a}{d+a} \end{bmatrix} \qquad \{3.18\}$$

Substituting into {3.7} yields

$$y_k = \frac{y_0[\, a + d + (a - d)k\,] + 2bk}{y_0 2ck + a + d + (d - a)k} \qquad \{3.19\}$$

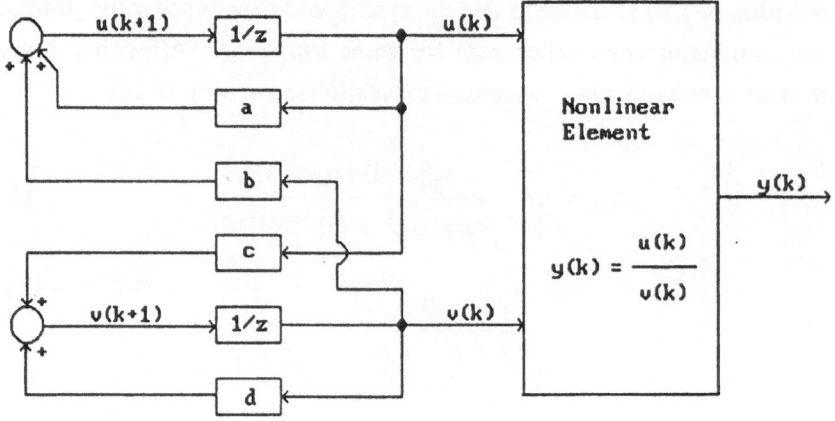

Figure 3.1

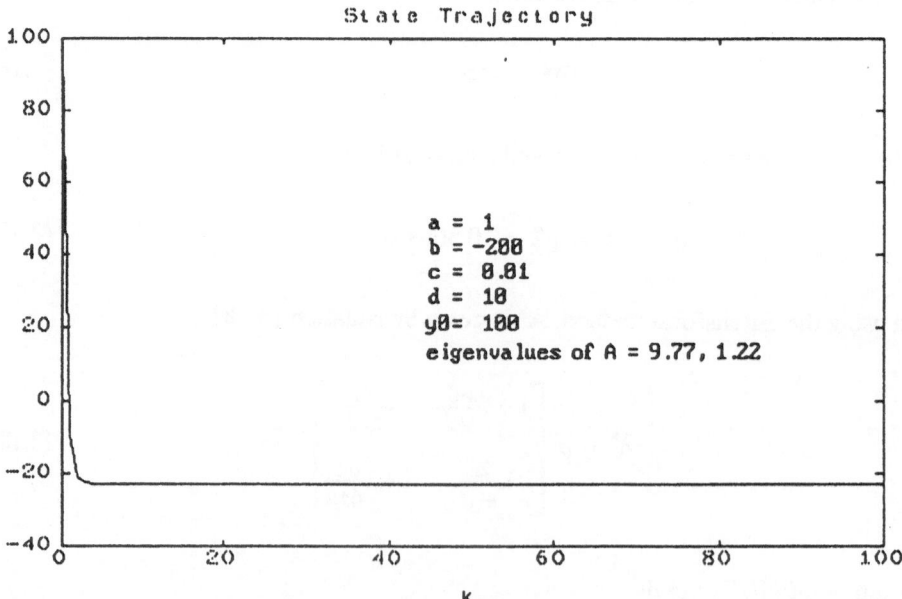

Figure 3.2

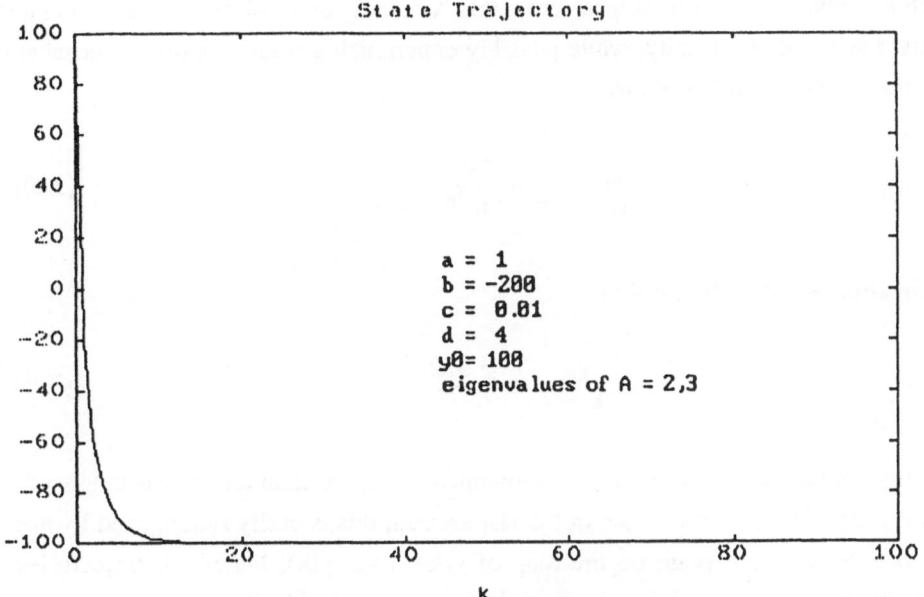

Figure 3.3

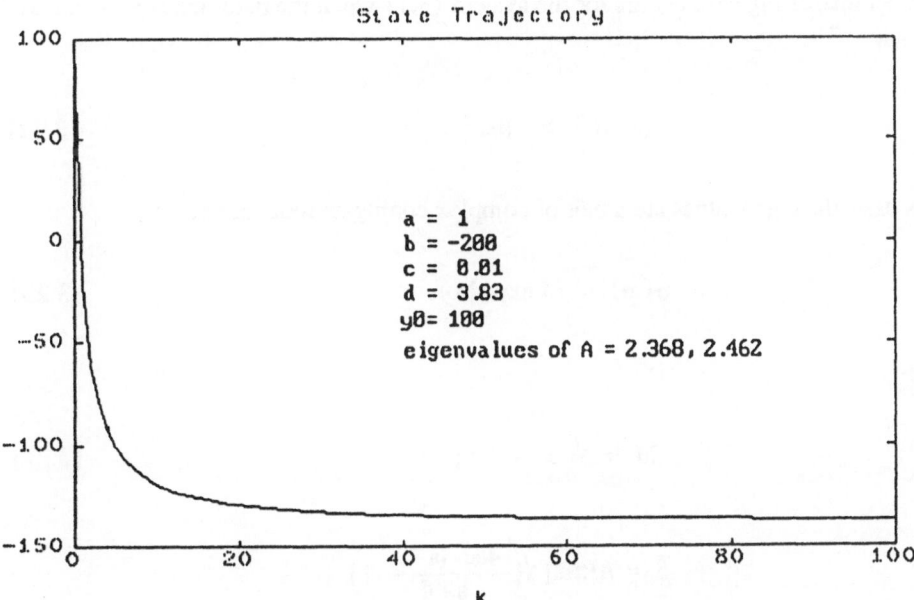

Figure 3.4

Much like that of the "overdamped" case, this "critically damped" case will tend to a finite constant as k goes to infinity, while possibly experiencing some "transient" terms at first. The steady-state value is given by

$$\lim_{k \to \infty} y_k = \frac{y_0(a - d) + 2b}{y_0 2c + d - a} \tag{3.20}$$

which, after use of {3.16}, yields

$$\lim_{k \to \infty} y_k = \frac{a - d}{2c} \tag{3.21}$$

Note that, in both this case and the "overdamped" case, the final attractor is independent of the value of y(0). As was shown in the last section, this is easily rationalized by noticing only one stable fixed point on the map of y(k+1) vs. y(k); hence, all trajectories will eventually converge to it. Some sample trajectories of the "critically damped" case are shown in Figures 3.5 and 3.6.

(3) Complex Eigenvalues

The most interesting case occurs for the system {3.1} when the parameters a, b, c, and d are such that

$$-(a - d)^2 > 4bc \tag{3.22}$$

In this case, the eigenvalues are a pair of complex conjugate roots of values

$$p1, p1 = M \exp(\pm j\beta) \tag{3.23}$$

where

$$M = \sqrt{[ad - bc]} \tag{3.24}$$

$$\beta = Arctan\left\{ \sqrt{\left[\frac{4bc - (a - d)^2}{a + d}\right]} \right\}$$

The matrix trajectory A^k and the system trajectory are as follows:

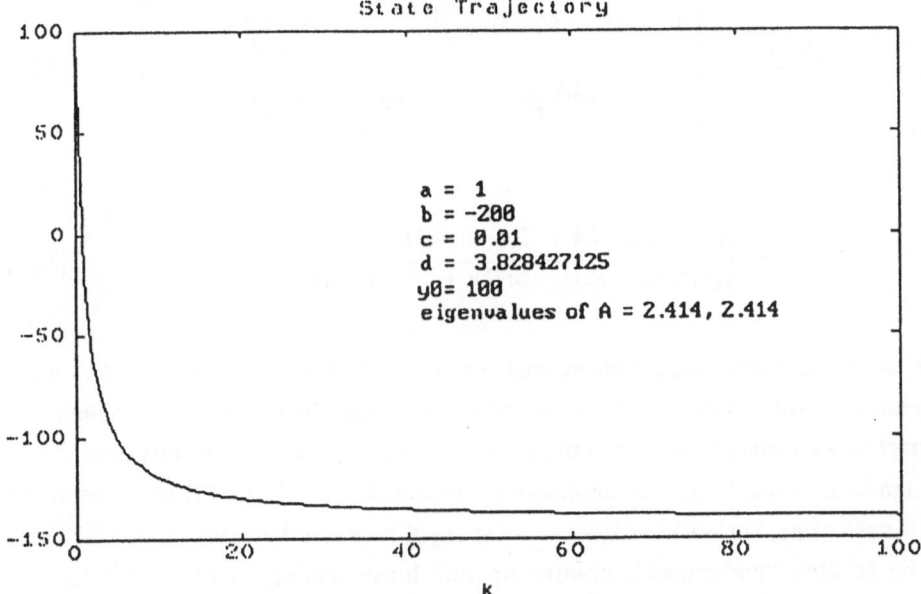

Figure 3.5

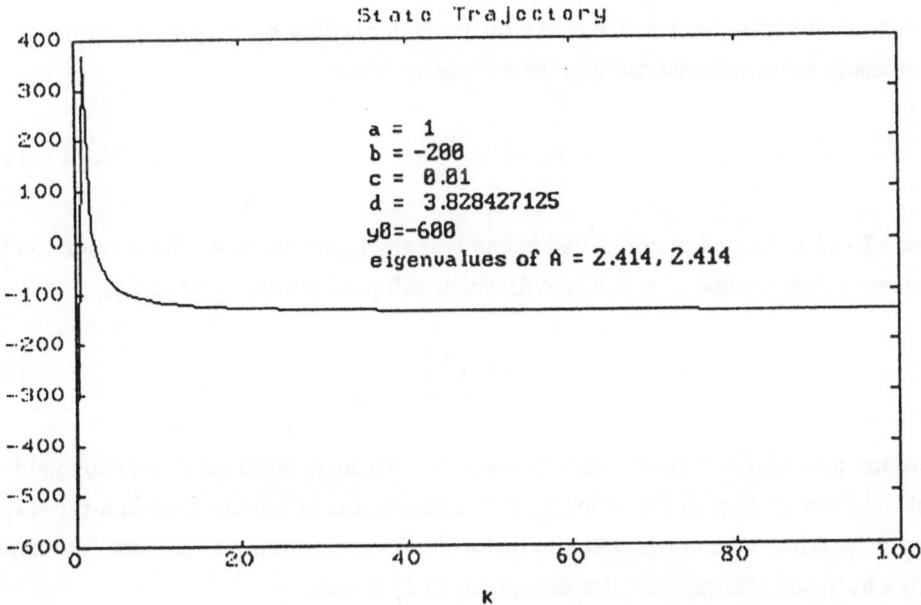

Figure 3.6

$$A^k = M^{k-1} \begin{bmatrix} M\cos(k\beta) + .5(a-d)\dfrac{\sin k\beta}{\sin \beta} & b\dfrac{\sin k\beta}{\sin \beta} \\ c\dfrac{\sin k\beta}{\sin \beta} & M\cos(k\beta) + .5(d-a)\dfrac{\sin k\beta}{\sin \beta} \end{bmatrix} \qquad \{3.25\}$$

$$y_k = \frac{y_0[\, M\sin\beta\cos k\beta + .5(a-d)\sin k\beta\,] + b\sin k\beta}{y_0 c\sin k\beta + M\sin\beta\cos k\beta + .5(d-a)\sin k\beta} \qquad \{3.26\}$$

There are several interesting comments that can be made about the system trajectory when the eigenvalues of A are complex. In the previous cases, the ratio of numerator to denominator approached a constant value, even though each could "blow up". In this case, however, although both numerator and denominator are bounded, the ratio $\{3.26\}$ never approaches a steady-state value, but wanders forever somewhere between the values $(-\infty, \infty)$. This case will be labeled "undamped", continuing our loose analogy with a standard linear second-order system.

For $\{3.26\}$ to be periodic we require

$$y_k = y_{k+N} \qquad\qquad \text{all k, natural N} \qquad\qquad \{3.27\}$$

Since there are both sines and cosines of terms depending on k in both numerator and denominator, the necessary condition for periodicity is that

$$k2 - k1 = \frac{2\pi N}{\beta} \qquad\qquad \{3.28\}$$

where $k2 - k1$ is the period of oscillation and N is any natural number. This condition leads to values of ß (a function of a, b, c, and d) which will yield periodic trajectories:

$$\beta = \pi\,\frac{m1}{m2} \qquad\qquad \{3.29\}$$

where m1 and m2 are natural numbers $m2 \geq m1$. Clearly, although it is uncountable, the number of sets (a, b, c, d) for which $\{3.29\}$ is satisfied is of volume zero in a-b-c-d space. Loosely speaking, then, the probability that a given set of parameters (chosen "at random") yields a truly periodic trajectory for the system $\{3.1\}$ is zero.

In a practical sense the above arguments mean that for "practically all" sets of parameters a, b, c, and d, the trajectory of the system will be either aperiodic or will be of a period so large that it looks aperiodic. However, since many ß will be of the form

$$\beta \approx \pi\frac{m1}{m2}$$

then many trajectories will look like periodic ones which slowly slip away from their apparent attractors over time.

Shown in Figures 3.7 through 3.9 are some of the more interesting sample trajectories associated with the "oscillatory" case. It is encouraging to note how the results obtained in this section are consistent with the observations of Chapter 2. The most important of these are the conditions for periodicity. In Chapter 2, equation {2.11} was the derived condition for oscillatory behavior; in this section, this corresponds to imaginary eigenvalues as dictated by equation {3.11}.

Furthermore, for periodic oscillations of period 3 or greater, we see from equation {3.24} that

$$\tan\{\frac{2\pi}{N}\} = \sqrt{[\frac{-4bc - (a-d)^2}{a+d}]} \qquad \{3.30\}$$

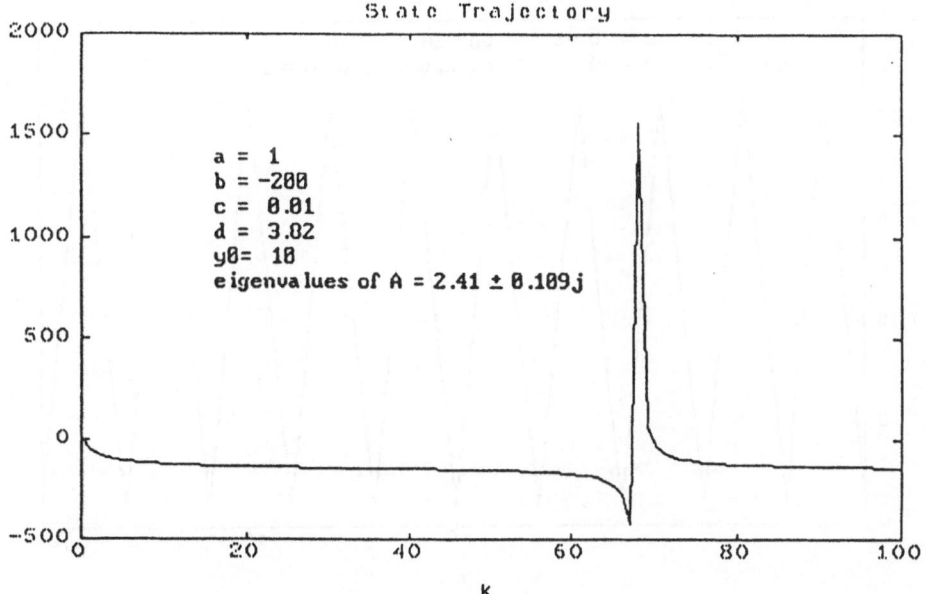

Figure 3.7

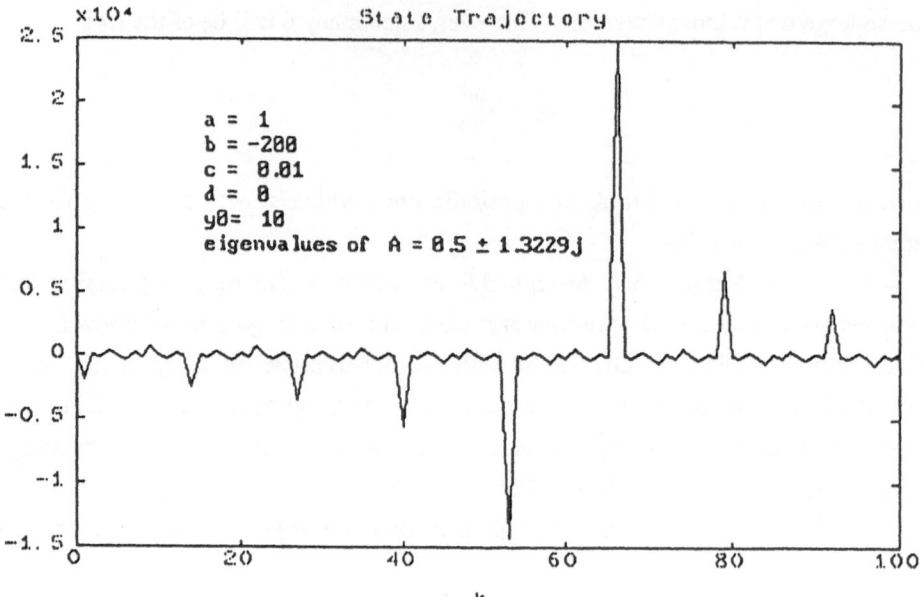

Figure 3.8

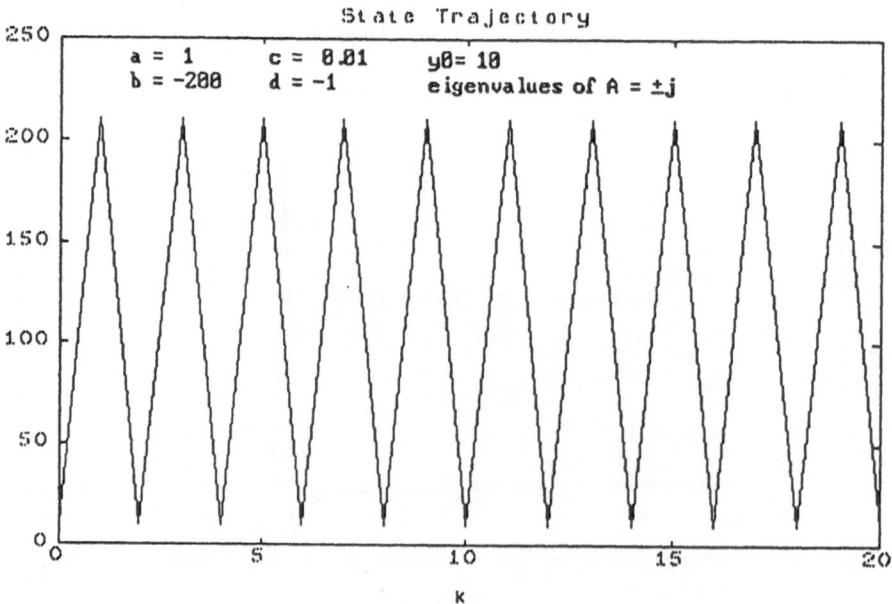

Figure 3.9

is the necessary condition for period N, assuming equation {3.22} is satisfied. Looking back at Chapter 2, equation {2.16} was constructed as necessary for period 3 behavior, and it is indeed seen to be {3.30} with N = 3.

Finally, the peculiar case of period 2 behavior when a = -d is also explained by the results of this section. When

$$(a-d)^2 + bc < 0$$

we obtain ß = π/2 in equation {3.24} for a = -d. Equation {3.26} then yields

$$y_k = \frac{y_0[\, M\cos(k\pi/2) + a\sin(k\pi/2)\,] + b\sin(k\pi/2)}{y_0 c\sin(k\pi/2) + M\cos(k\pi/2) + d\sin(k\pi/2)} \qquad \{3.31\}$$

which oscillates between the values

$$y_k = y_0, \ \frac{y_0\, a + b}{y_0\, c + d}, \ y_0, \ \frac{y_0\, a + b}{y_0\, c + d}, \ y_0, \ \cdots$$

In addition, when

$$(a-d)^2 + bc > 0$$

then equation {3.13} can also be shown to degenerate to this period 2 sequence.

To summarize briefly the last two chapters, we have taken the nonlinear recursion equation

$$y_{k+1} = \frac{a y_k + b}{c y_k + d} \qquad \{3.1\}$$

and have analyzed it both qualitatively and quantitatively. First, through a look at the map of y(k+1) vs. y(k) we intuitively predicted and explained the basic trajectory types; in addition, it was shown (without rigorous proof) that the system is deterministic and is not chaotic.

We then proceeded to perform a rigorous derivation of the closed-form solution to the one-dimensional trajectory of equation {2.1}, which was then shown to yield results identical to the intuitive explanations of Chapter 2. This one-dimensional analysis was very useful in terms of clarity and ease of derivation. In addition, an in-depth study of this case was important because it is these same mechanisms which induce similar behavior in multidimensional systems.

A general extension of the derivation in this section to the multidimensional case as well as applications are included in the following two chapter appendices.

CHAPTER THREE, APPENDIX A
EXTENSION TO THE MULTIDIMENSIONAL CASE

This section considers the n-dimensional counterpart of equation {3.1}, which is

$$Y_{k+1} = [AY_k + B][CY_k + D]^{-1} \qquad \{3.a.1\}$$

where $Y(k)$ is n-square, and the matrices A, B, C, and D are n-square and, when necessary, invertible.

Emulating the one-dimensional case, we consider $Y(k)$ to be the product of two matrices $U(k)$ and $V(k)^{-1}$:

$$Y_k = U_k V_k^{-1} \qquad \{3.a.2\}$$

where $U(k)$ and $V(k)$ are n-square and $V(k)$ is invertible.
The following manipulations are then performed on {3.a.1}:

$$U_{k+1}V_{k+1}^{-1} = [AU_kV_k^{-1} + B][CU_kV_k^{-1} + D]^{-1}$$

$$U_{k+1}V_{k+1}^{-1} = [AU_kV_k^{-1} + B] I [CU_kV_k^{-1} + D]^{-1}$$

$$U_{k+1}V_{k+1}^{-1} = [AU_kV_k^{-1} + B] V_kV_k^{-1} [CU_kV_k^{-1} + D]^{-1}$$

$$U_{k+1}V_{k+1}^{-1} = [AU_k + BV_k][(CU_kV_k^{-1} + D)V_k]^{-1}$$

$$U_{k+1}V_{k+1}^{-1} = [AU_k + BV_k][CU_k + DV_k]^{-1} \qquad \{3.a.3\}$$

Comparing the left and right-hand sides of equation {3.a.3} yields

$$U_{k+1} = AU_k + BV_k$$
$$V_{k+1} = CU_k + DV_k \qquad \{3.a.4\}$$

which can be written in the partitioned matrix form

$$\begin{bmatrix} U_{k+1} \\ V_{k+1} \end{bmatrix} = \begin{bmatrix} A & B \\ C & D \end{bmatrix} \begin{bmatrix} U_k \\ V_k \end{bmatrix} \qquad \{3.a.5\}$$

$$\text{(2n x n)} \qquad \text{(2n x 2n)} \ \text{(2n x n)}$$

where $Y(k) = U(k)V(k)^{-1}$ for all k. Again we see that the problem has been broken down into a linear 2n-dimensional one, where the output is a nonlinear function of the states which propagate linearly in time. Although any $U(0)$ and $V(0)$ which satisfy

$$Y_0 = U_0 V_0^{-1}$$

would work, we arbitrarily choose $V(0) = I$ for simplicity. Hence, $Y(k)$ can be computed in closed form as

$$Y_k = U_k V_k^{-1} \quad : \quad U_0 = Y_0 \ ; \ V_0 = I \qquad \{3.a.6\}$$

where

$$\begin{bmatrix} U_k \\ V_k \end{bmatrix} = \begin{bmatrix} A & B \\ C & D \end{bmatrix}^k \begin{bmatrix} Y_0 \\ I \end{bmatrix}$$

$$= \ G^k \begin{bmatrix} Y_0 \\ I \end{bmatrix} \qquad \{3.a.7\}$$

with

$$G^k \ \text{also inverse z-transform of} \ [\ z(zI - G)^k \ ] \qquad \{3.a.8\}$$

The analysis of the types of trajectories followed by the elements of Y(k) also follows very closely the analysis for the one-dimensional case. It does not suffice to determine only the behavior of G^k; it is also necessary to find how the expression

$$U_k V_k^{-1} = [G11_k Y_0 + G12_k][G21_k Y_0 + G22_k]^{-1} \qquad \{3.a.9\}$$

propagates in time, where

$$G^k = \begin{bmatrix} A & B \\ C & D \end{bmatrix}^k = \begin{bmatrix} G11_k & G12_k \\ G21_k & G22_k \end{bmatrix} \qquad \{3.a.10\}$$

Thus, the eigenvalues of G do not have to be inside the unit circle for Y(k) = $U(k)V(k)^{-1}$ to follow a bounded trajectory. It is the *type* (e.g. period 1 attractor, oscillatory) rather than the magnitude of trajectory followed by G^k which is important in determining how Y(k) behaves.

Unlike the one-dimensional case, analytical expressions for G^k in terms of the partitions A, B, C, and D would be difficult to derive (the literature reviewed for this purpose yielded no general methods). Although we could tediously find equation {3.a.7} in terms of the individual constituents of the partitions, this would have to be done for a specific dimension and not for the "n"-dimensional case. Most importantly, no new insights would be gained since, as will be shown for the linear quadratic regulator in Chapter 5, no new behaviors arise which differ from the one-dimensional case. This, of course, is to be expected, since equation {3.a.1} simply cannot display chaotic behavior, which is the only fundamentally new behavior that can arise.

The remainder of the investigation of this system will be performed under the auspices of the linear quadratic regulator in Chapter 5.

CHAPTER THREE, APPENDIX B
SOME INTERESTING APPLICATIONS OF
THE HYPERBOLIC MAP

This section briefly introduces some physical cases where the one-dimensional equation {2.1} arises. Four applications are discussed: (1) an iterative square-root finding algorithm, (2) real image propagation for two inward-facing parabolic mirrors, (3) input resistance of cascaded resistance networks, and (4) input reactance of cascaded L-C networks.

(1) Square Root Algorithm

Consider the following specific case of equation {2.1}:

$$y_{k+1} = \frac{y_k + P}{y_k + 1} \qquad \{3.b.1\}$$

This is equation {2.1} with $a = c = d = 1$ and $b = P$. Using the methods of Chapter 3, it is seen that {3.b.1} is an iterative method of finding the square root of the real number P, and the only stipulation is that $P \geq 0$.

Using {3.11} to find the system poles p1 and p2 yields

$$p1, p2 = 1 \pm \sqrt{P} \qquad \{3.b.2\}$$

If $P > 0$, then it is guaranteed that the system trajectory will approach a *single* fixed point in the steady state. Furthermore, this value, y(SS), is found by using {3.15} which yields

$$y_{ss} = \sqrt{P} \qquad \{3.b.3\}$$

It is encouraging to note further that *any* initial guess y(0) will converge to $\sqrt{P}$, as the results of Chapter 3 guarantee. Finally, we note that if P = 0, the equation will, by {3.21}, converge to y(SS)=0 as desired.

(2) Inward-Facing Parabolic Mirrors

Consider two inward-facing ideal parabolic mirrors of equal focal length f separated by a distance a', as depicted in Figure 3.b.1. An object (let it be a candle), when placed between the mirrors, serves as an input to, say, mirror #1, and an inverted real image results. This image then serves as an input to mirror #2; the resulting output image is fed back into mirror #1, and so on, back and forth. (This assumes, of course, that the candle is put in an appropriate place for such behavior to happen.) In addition, since nothing is special about mirror #1, there is also a "candle - #2 - #1 - #2 - ..." image propagation in addition to this "candle - #1 - #2 - #1 - ..." propagation. Without loss of generality, let us focus on only one of these image propagations, since they are just complements of each other.

Clearly, there is an established iterative pattern, each value of "k" leaving behind a real image. The location to which these images converge is dictated by an equation of the form of {2.1}, as is now shown. Let d represent the object distance from the input mirror and let d' represent the corresponding image distance. Clearly, for the kth and (k+1)st iterations,

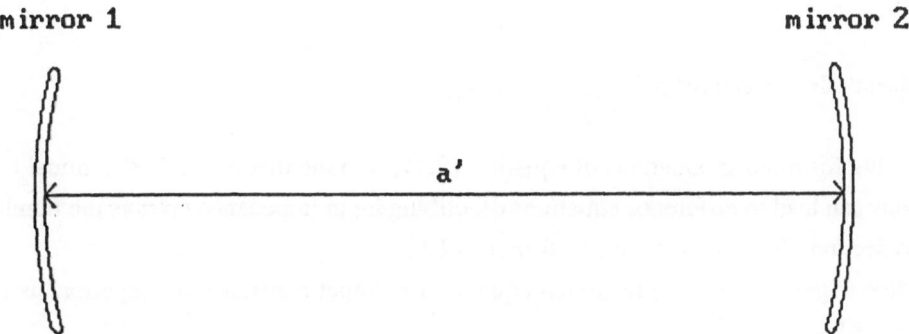

mirror 1   mirror 2

a'

Figure 3.B.1

$$d'_k = a' - d_{k+1} \qquad \{3.b.4\}$$

because the kth image is the object for the (k+1)st iteration. From geometrical optics it is known that

$$\frac{1}{f} = \frac{1}{d'_k} + \frac{1}{d_k} \qquad \{3.b.5\}$$

which, after substituting {3.b.4} yields

$$d_{k+1} = \frac{(a' - f)d_k + a'f}{d_k - f} \qquad \{3.b.6\}$$

Thus, the propagation of real images is dictated by an equation of the form of {2.1}, with $a = a' - f$, $b = -a'f$, $c = 1$, and $d = -f$. Expecting no "intermittent" behavior of *real* image propagation, we use equation {3.10} to find the values a' for convergent behavior, which yields the result

$$a' > 4f \qquad \{3.b.7\}$$

Equation {3.b.7} serves simply to verify our physical intuition concerning the mirrors, for if the mirrors were less than 4 focal lengths apart, then *virtual* images (corresponding to negative d's) would result; when the images are *real* and actually exist in space, there is no strange behavior.

(3) Cascaded Resistors

Searching for more applications of equation {2.1}, we note that cascading identical circuit sections can lead to difference equations describing input impedance (versus the number of added sections "k") which are of the form of {2.1}.

Consider Figure 3.b.2. The recursion equation for "input resistance propagation" is given by {3.b.8},

$$R_{k+1} = \frac{RR_k + a'R^2}{R_k + (1+a')R} \qquad \{3.b.8\}$$

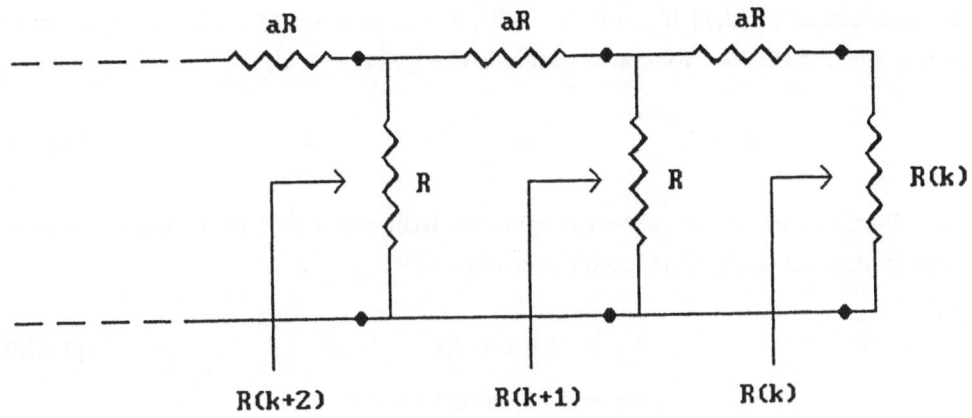

Figure 3.B.2

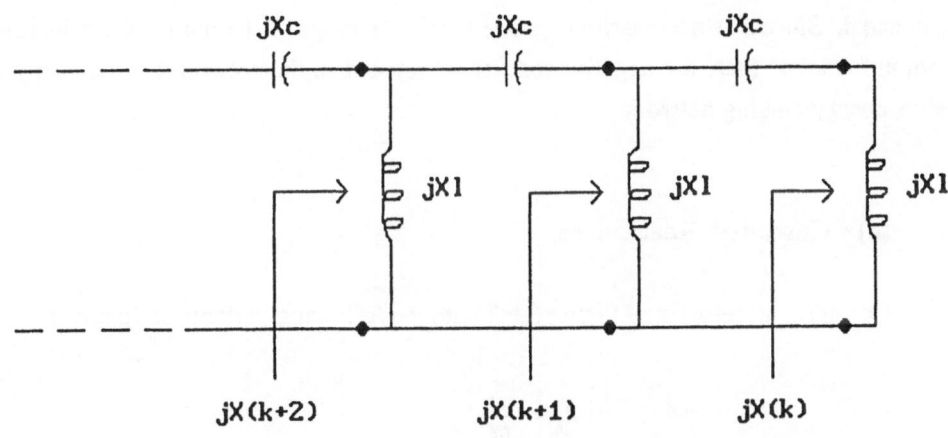

Figure 3.B.3

which is identical to {2.1} if $a = R$, $b = a'R^2$, $c = 1$, and $d = (1 + a')R$. Again invoking {3.10} to find the values a' for convergent behavior quickly yields

$$a' > 0 \qquad \text{or} \qquad a' < 4 \qquad \{3.b.9\}$$

Furthermore, the steady-state value for R(k) when a' is in the region satisfying {3.b.9} is obtained using {3.15}, which readily yields

$$R_{SS} = .5R\{ -a' + \sqrt{[a'^2 + 4a']} \} \qquad \{3.b.10\}$$

(plus for a' > 0, minus for a' < -4)

It is noteworthy that R(SS) is always positive if a' > 0 and is always positive if a' < -4.

If a' is between -4 and 0, the input resistance never converges to a constant steady-state value as more and more sections are added; the dynamics of the values R(k) takes on as k increases is dictated by {3.26} with the aforementioned substitutions made for a, b, c, and d. Since a' is necessarily negative for this "strange" behavior to occur, its causes are not unfounded: there are negative resistances present so the network is no longer just a passive, energy-sinking network.

(4) Cascaded Reactances

Consider the network of Figure 3.b.3 with the following variable assignments:

$$X_C = \frac{-1}{\omega C}$$

$$X_L = \omega L$$

Furthermore, assign $X_C = a'X_L$ (with a' < 0 necessarily).

Then it is easily shown that, if X(k) denotes the input reactance of the entire cascaded network,

$$X_{k+1} = \frac{X_L X_k + a' X_L^2}{X_k + (a' + 1)X_L} \qquad \{3.b.11\}$$

Clearly, this equation is identical to {3.b.8}, except with $R = X_L$, and is subject to the same dynamics for $X(k)$ as {3.b.8} is for $R(k)$. The only difference is that a' must by necessity be less than zero for this circuit. For a' < -4, the input reactance approaches a constant steady-state value X(SS), while for a' in the range [-4,0] $X(k)$ is intermittent, never approaching a constant value.

In this appendix we have shown some places where the mapping {2.1} arises and have used the results of Chapter 3 to predict their behaviors in varying parameter regions. The next two sections turn to our primary reason for studying {2.1}, the discrete-time Linear Quadratic Regulator.

CHAPTER 4

THE LINEAR QUADRATIC REGULATOR - BACKGROUND

As stated in Chapter 1, a prime motivation for studying the hyperbolic map was to analyze the dynamics of the discrete linear quadratic regulator (LQR) when certain parameters, specifically the performance index parameters, are assigned "abnormal" values: what will happen to the system, why, and could it be useful?

Included first for completeness and understanding of the results is a brief and intuitive derivation of the linear quadratic regulator solution. The heart of the solution is a recursive algorithm which is an n-dimensional discrete-time Riccati equation. A "typical" one-dimensional LQR is then displayed for reference purposes, followed by a discussion of necessary conditions for a steady-state solution to the discrete-time Riccati equation. Finally, an optimal root-locus technique is mentioned for the very practical case of time-invariant sub-optimal LQR feedback. The discussion in general is geared toward Chapter 5, which will examine the effects of "negative Q".

The following discussion follows very closely that of Lewis (1986).

Any regulator system is, by definition, a special case of a tracking system; specifically, the value of zero is tracked. It is often desired to accomplish this with a minimum of input energy, intermediate-state squared error, and final-state squared error (between the actual and desired performance). Clearly, this implies the necessity of creating some quadratic performance measure which we would like to minimize for all points on the trajectory as the states are forced to zero. The optimal control strategy for a system would be one which causes its states to tend to zero while keeping this quadratic performance measure to a minimum.

Let the n-dimensional system with "m" inputs be represented as shown in Figure 4.1. Consider the following definition for a performance measure J:

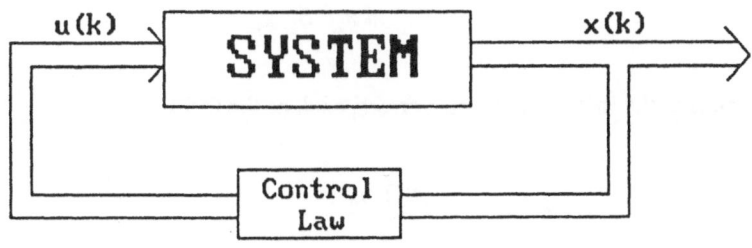

SYSTEM : $x(k+1) = A(k) * x(k) + B(k) * u(k)$

where $x(k)$ is $n \times 1$
             $u(k)$ is $m \times 1$
             $A(k)$ is $n \times n$
             $B(k)$ is $n \times m$

Figure 4.1

$$J_i = .5\, x_N^T S_N x_N + .5 \sum_{k=i}^{N-1} \left[\, x_k^T Q_k x_k + u_k^T R_k u_k \, \right] \tag{4.1}$$

where

       i = starting time of control law application

       N = final time of control law application

       x_k = state vector at intermediate time k

       x_N = state vector at final time N

       u_k = input vector at intermediate time k

       u_N = input vector at final time

       S_N = final state weighting matrix, n x n, symmetric, positive semidefinite

       Q_k = intermediate state weighting matrix, n x n, symmetric, positive semidefinite

       R_k = input energy weighting matrix, n x n, symmetric, positive semidefinite

J(i), then, is a hybrid sum of "everything which we would like to be small", with some weighting matrices included for adjustability: S(N) for the final state squared error, which is equal to

$$.5 \, x_N^T \, S_N \, x_N \, ,$$

Q(k) for the sum of the intermediate state errors, each of the form

$$.5 \, x_k^T \, Q_k \, x_k \, ,$$

and R(k) for the sum of the input energies at each time k,

$$.5 \, u_k^T \, R_k \, u_k \, .$$

The key to the problem is the realization that we have some function J(i) which we want to be minimized, and we also have some constraints to satisfy, specifically, the system dynamics equation itself,

$$x_{k+1} \; = \; A_k \, x_k \; + \; B_k \, u_k \qquad\qquad \{4.2\}$$

for all times k in the interval [i,N].

Such a problem points directly to solution by the Lagrange multiplier method, only in a more complicated form. Instead of one Lagrange multiplier, we have an n x 1 vector of them (this vector we shall call λ, the "costate" of the system). In addition, since J(i) involves a summation from k = i

to N, these costates will be time-varying. The net result from using such a technique will be

a set of recursion formulae invloving x(k), λ(k), u(k), and a "cost kernel matrix" S(k) (a.k.a. "performance index kernel sequence") which, when satisfied, will yield the input u(k) and the trajectory x(k) that minimizes J(i) for the chosen S(N), Q(k), and R(k).

The Lagrange multiplier method, in general, involves building a function

$$J' \; = \; (\text{expression to be minimized}) \; + \; \lambda(\text{the constraint}) \qquad\qquad \{4.3\}$$

and then setting the total differential of J' (with respect to the expression variables and to λ) equal to zero. *Note that this will find a critical point of the cost function J(i), not necessaily a minimum.* For our purposes here, however, we assume we have found an absolute minimum; this should always be the case when S(N), Q(k), and R(k) are positive semidefinite for all k.

We shall not wade through the details of the solution but, in summary, it involves taking the total differential of J' with respect to $\lambda(k)$, $x(k)$, $u(k)$, and $x(N)$, and setting it equal to zero:

$$\partial J' = \{ \quad \}\partial\lambda_k + \{ \quad \}\partial x_k + \{ \quad \}\partial u_k + \{ \quad \}\partial x_i + \{ \quad \}\partial x_N = 0 \qquad \{4.4\}$$

Loosely speaking, each of the above braced expressions results in a necessary equation which must always be satisfied. The first three differentials (J' with respect to $\lambda(k)$, $x(k)$, and $u(k)$) yield the following equations:

$$\text{"state equation"} \quad x_{k+1} = A_k x_k + B_k u_k \qquad \{4.5\}$$

$$\text{"costate equation"} \quad \lambda_{k+1} = Q_k x_k + A_k^T \lambda_{k+1} \qquad \{4.6\}$$

$$\text{"stationarity condition"} \quad 0 = R_k u_k + B_k^T \lambda_{k+1} \qquad \{4.7\}$$

Then, the last two differentials (J' with respect to x(i) and x(N)) yield two boundary conditions. However, only the latter one with respect to x(N) is of consequence, because the state initial condition x(i) is unchangeable and so $\partial x(i)$ must be zero anyway. This consequential boundary condition is

$$\left[\frac{\partial}{\partial x_N}\left[.5\, x_N^T S_N x_N \right] - \lambda_N \right]\partial x_N = 0 \qquad \{4.8\}$$

If we choose to fix the final state x(N) to zero by design (force the states to be some exact predetermined value x(N) at time N) then equation {4.7} is automatically satisfied, since $\partial x(N) = 0$. On the other hand, if we let the final state be free (i.e. if we don't restrict it to any exact value at time k = N), then equation {4.8} yields

$$\lambda_N = S_N x_N \qquad \{4.9\}$$

Concerning the former case $\partial x(n) = 0$, it can be shown that the restriction of the final state leads to an open-loop optimal control strategy. Such open-loop strategies are not robust against system parameter changes, noise, etc., so we shall not deal with this case; rather, we shall pursue the other case where x(N) is free, which leads to a closed-loop optimal control strategy.

To find the necessary closed-loop control strategy, we make the assumption that the costate,
$\lambda(k)$, is a direct function of the state, $x(k)$, through the relation

$$\lambda_k = S_k x_k \qquad k = [i,N] \qquad \{4.10\}$$

where $S(k)$ is called the "cost kernel matrix" as mentioned previously. If this cost kernel
matrix can be found for $k = [i,N]$ then $\lambda(k)$, and hence $u(k)$ (from $\{4.7\}$), can be found in
terms of $x(k)$, which is our closed-loop state feedback goal. Solving for $S(k)$ is rather
simple:

From $\{4.7\}$, $\{4.10\}$:

$$u_k = -R_k^{-1} B_k^T \lambda_{k+1} = -R_k^{-1} B_k^T S_{k+1} x_{k+1} \qquad \{4.11\}$$

From $\{4.5\}$, $\{4.11\}$:

$$x_{k+1} = A_k x_k - B_k R_k^{-1} B_k^T S_{k+1} x_{k+1} \qquad \{4.12\}$$

$$x_{k+1} = \left[I + B_k R_k^{-1} B_k^T S_{k+1} \right]^{-1} A_k x_k \qquad \{4.13\}$$

From $\{4.10\}$, $\{4.6\}$, $\{4.13\}$:

$$S_k x_k = Q_k x_k + \left[A_k^T S_{k+1} \left[I + B_k R_k^{-1} B_k^T S_{k+1} \right]^{-1} A_k \right] x(k) \qquad \{4.14\}$$

Cancelling the $x(k)$ on both sides yields:

$$S_k = Q_k + A_k^T S_{k+1} \left[I + B_k R_k^{-1} B_k^T S_{k+1} \right]^{-1} A_k \qquad \{4.15\}$$

Equation $\{4.15\}$ is the discrete-time Riccati equation; it is easily converted to the
form of $\{3.a.1\}$, the multidimensional hyperbolic map.

Starting at $k = N$, $S(k)$ can be calculated off-line, because no parameters of $S(k)$
depend on the system trajectory $x(k)$. From $\{4.5\}$ and $\{4.11\}$, then,

$$u_k = -R_k^{-1} B_k^T S_{k+1} \left[A_k x_k + B_k u_k \right] \tag{4.16}$$

or

$$u_k = -\left[B_k^T S_{k+1} B_k + R_k \right]^{-1} B_k^T S_{k+1} A_k x_k \tag{4.17}$$

Equation {4.17}, then, is the optimal control law for the system which, while time-varying, is nevertheless a deterministic feedback of the system states. The negative of the coefficient of x(k) in {4.17} is called the Kalman gain K(k); {4.17} is then cast as

$$u_k = -K_k x_k \tag{4.18}$$

Summarizing, then, given a system to regulate

$$x_{k+1} = A_k x_k + B_k u_k \tag{4.2}$$

subject to the performance criterion

$$J_i = .5 x_N^T S_N x_N + .5 \sum_{k=i}^{N-1} \left[x_k^T Q_k x_k + u_k^T R_k u_k \right] \tag{4.1}$$

a closed-loop optimal control law can be realized by

(1) Iterating equation {4.15} *backward* in time from k = N to k = i, with "initial condition" S(k) = S(N) for k = N.

(2) Using equation {4.17} for the feedback gain u(k) = -K(k)x(k).

This process is outlined in Figure 4.2.

One result concerning the LQR system of Figure 4.2, included here without proof, concerns the physical meaning of the sequence x(k). It can be shown that, if J(k) represents the total cost to go before the final time N, then

$$J_k = .5 x_k^T S_k x_k \tag{4.19}$$

Thus, the total cost left to go is actually a known quantity at any given time k in [i,N]. This important result is referred to later in this chapter.

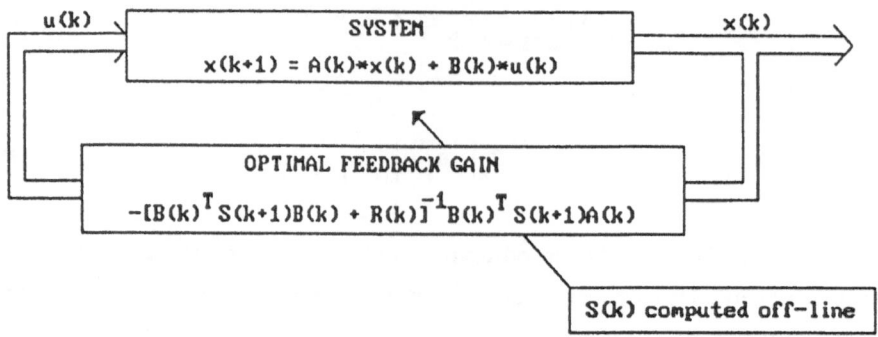

Figure 4.2

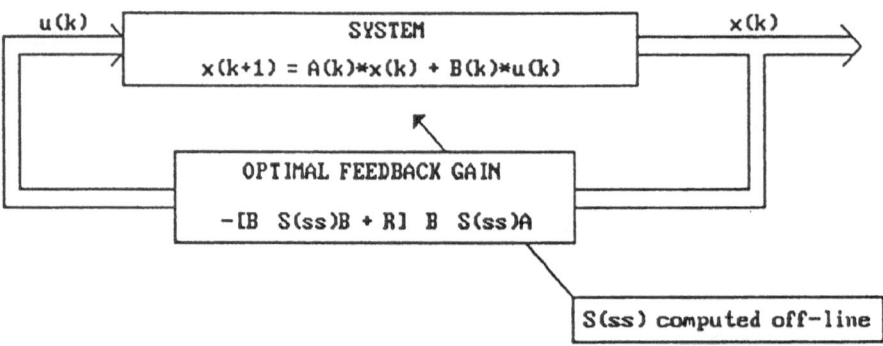

Figure 4.3

Let the system and performance parameters be time-invariant: $A(k) = A$, $B(k) = B$, $Q(k) = Q$, and $R(k) = R$. In most instances $S(k)$ will converge quickly to a constant attractor, $S(SS)$, when iterating backward in time. In general, then, this steady-state value can be used for computing $u(k)$ for *all* $k = [i,N]$, with very little change in he trajectory of $x(k)$. That is, $x(k)$ will go to zero in about the same amount of time, anyway, and the performance index will still be almost minimized.

If this is the case, then, we see that the value of N does not really matter, and may as well be infinity. The result is a *time-invariant* optimal feedback strategy which will work for all time, not just for $k = [i,N]$. Figure 4.3 diagrams this approach. A system such as Figure 4.3 is a "suboptimal feedback" LQR.

Figures 4.4 through 4.9 show a one-dimensional case of the LQR in action. The parameters used, which will be our reference set for the discussion of the one-dimensional case, are:

$$
\begin{aligned}
a &= 5 \\
b &= .2 \qquad\qquad \text{system parameters} \\
q &= 1 \\
r &= 1 \\
s_N &= 10 \qquad\quad\ \text{performance index parameters}
\end{aligned}
\qquad\qquad \{4.20\}
$$

The original system (a,b) is clearly unstable: the LQR, under normal operation, stabilizes the overall system, forcing the poles of the closed-loop system inside the unit circle. Note how similar the trajectory of x(k) is in Figures 4.6 and 4.9; in Figures 4.4 through 4.6 the exact time-varying feedback was used, while in Figures 4.7 through 4.9 the suboptimal feedback approach was used.

It is important to include here some further well-known results concerning the discrete LQR to assist in analyzing its behavior under abnormal performance index parameters (specifically, non-positive definite Q values). The first is the existence of closed-form solutions to the n-dimensional LQR's cost kernel matrix and, hence, system trajectory for any set of symmetric matrices Q, R, and S(N). The second is the existence of standardized root-locus and pole-placement methods for the easily implementable, sub-optimal, steady-state case of the LQR. Both of these results are developed and discussed fully in Lewis (1986) and are merely pointed to in this section.

First, we cast the equation {4.15} into the 2n-dimensional matrix recursion formula

$$
\begin{bmatrix} x_k \\ \lambda_k \end{bmatrix} = H \begin{bmatrix} x_{k+1} \\ \lambda_{k+1} \end{bmatrix}
\qquad\qquad \{4.21\}
$$

where

$$
H = \begin{bmatrix} A^{-1} & A^{-1}BR^{-1}B^T \\ QA^{-1} & A^T + QA^{-1}BR^{-1}B^T \end{bmatrix}
$$

and where, of course,

$$
\lambda_k = S_k x_k
\qquad\qquad \{4.10\}
$$

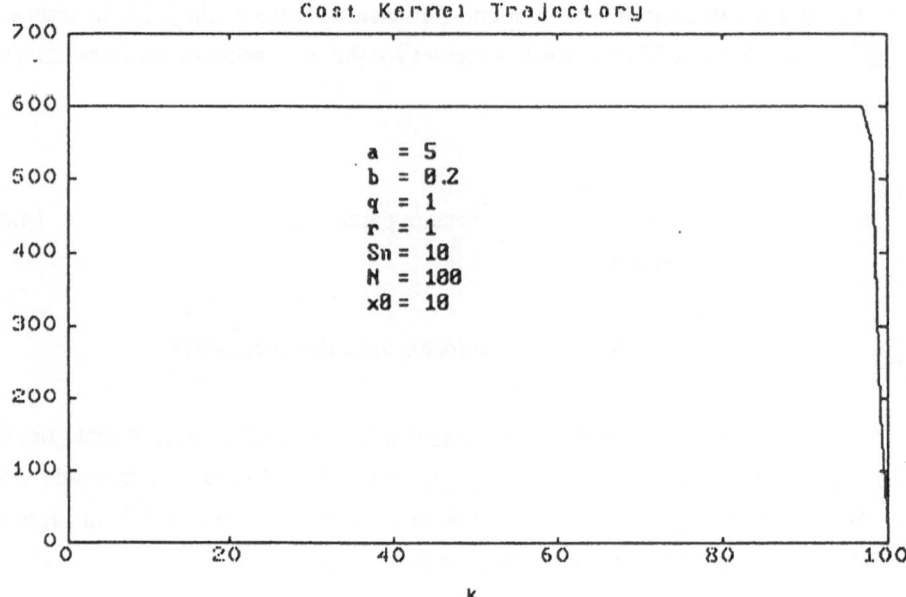

Figure 4.4

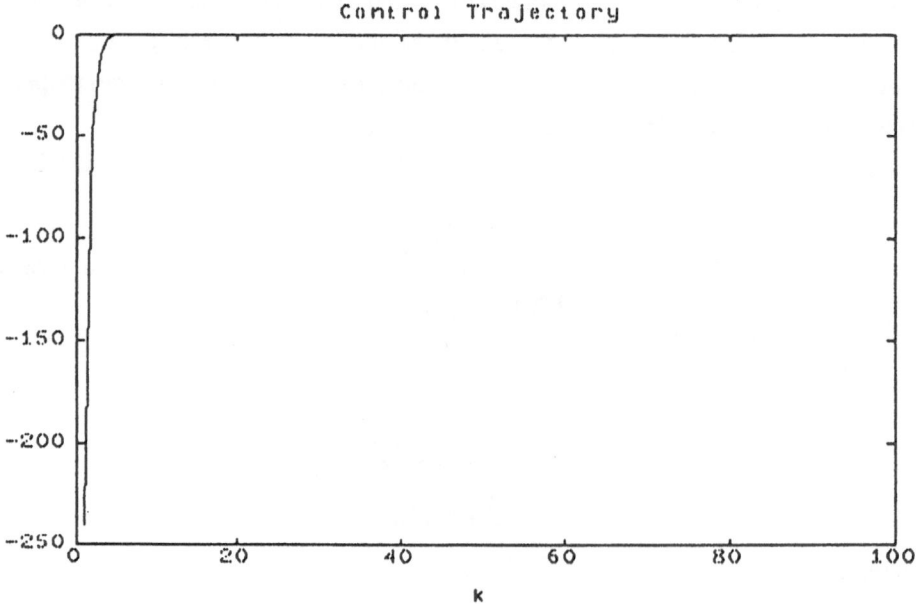

Figure 4.5

55

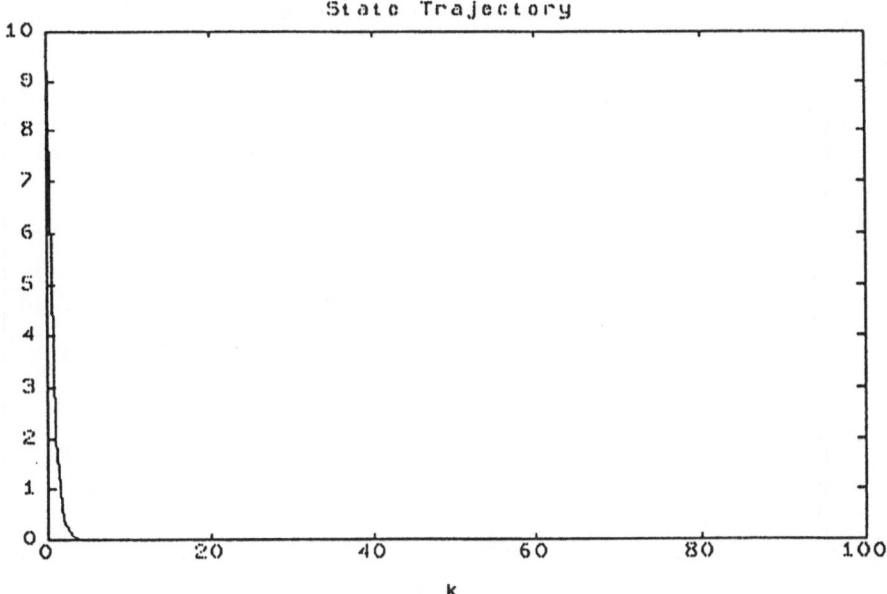

Figure 4.6

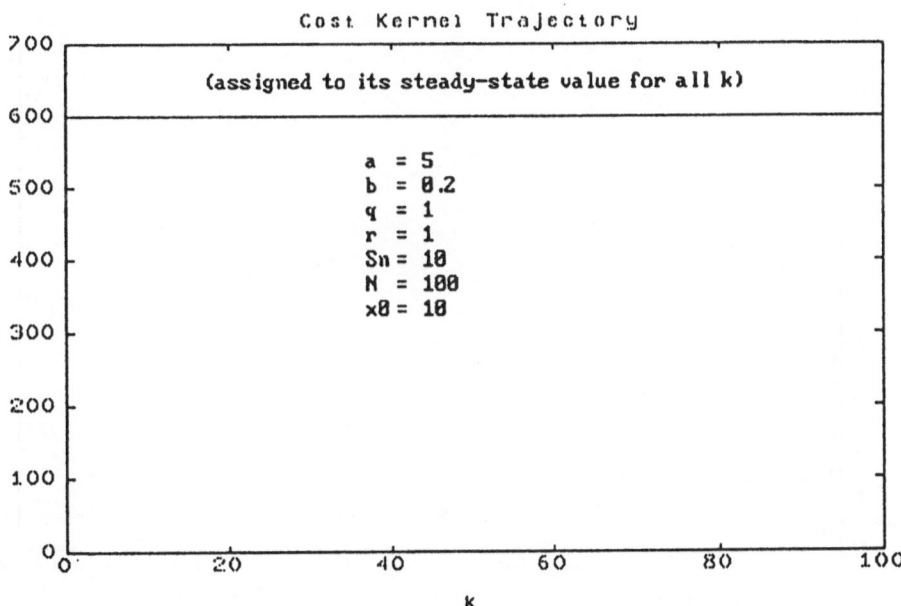

Figure 4.7

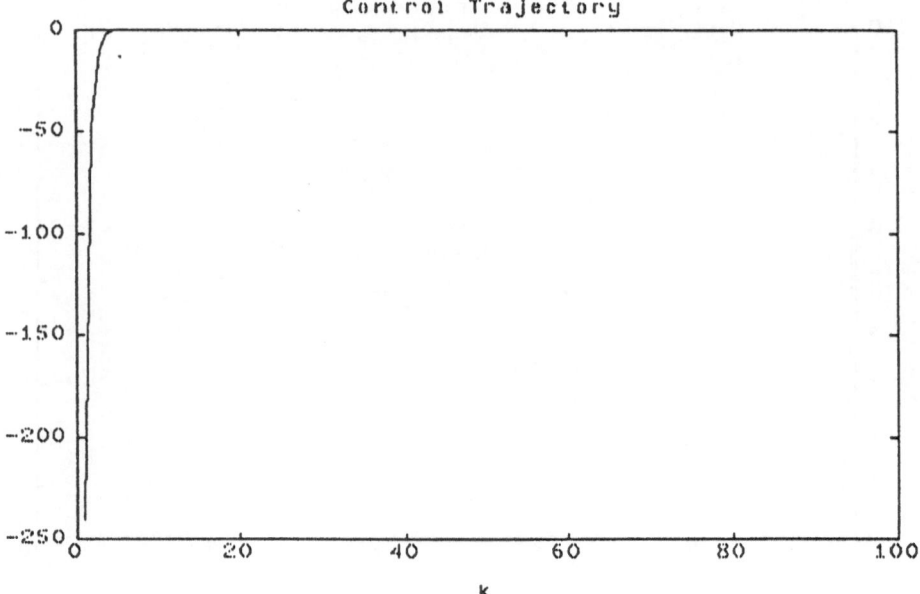

Figure 4.8

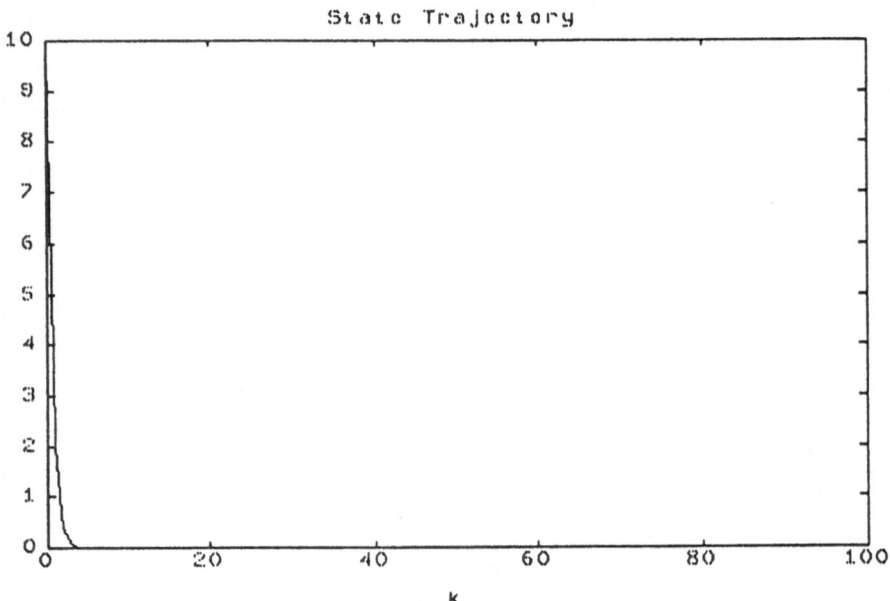

Figure 4.9

The key to finding S(k) analytically (as opposed to iteratively) for all k in [i,N] lies in the fact that H is "symplectic", that is, it satisfies

$$H^T \Gamma H = \Gamma \qquad \{4.22\}$$

where

$$\Gamma = \begin{bmatrix} 0 & I \\ -I & 0 \end{bmatrix}$$

Because of this, the "2n" eigenvalues of H are such that there are "n" stable eigenvalues and "n" unstable eigenvalues which are exact reciprocals of each other. From this, it can be shown that

$$S_k = \left[W21 + W22 \, T_k \right] \left[W11 + W12 \, T_k \right]^{-1} \qquad \{4.23\}$$

where

$$W_{2n \, x \, 2n} = \begin{bmatrix} W11 & W12 \\ W21 & W22 \end{bmatrix} \qquad \{4.24\}$$

is the matrix which diagonalizes H and where

$$T_k = -M^{-(N-k)} \left[W22 - S_N \, W12 \right]^{-1} \left[W21 - S_N \, W11 \right] M^{-(N-k)} \qquad \{4.25\}$$

with M being an n-diagonal matrix containing the "n" unstable eigenvalues of H. Note that this solution is independent of the nature of the symmetric matrices Q, R, and S(N) as long as H is diagonalizable.

When the eigenvalues of H are not on the unit circle, so that $M^{-(N-k)}$ will always approach zero under backward iteration, there will always exist a steady-state attractor S(SS). This is because T(k) in {4.25} will tend to zero under backward iteration, leaving

$$S_{SS} = W21 \, W11^{-1} \qquad \{4.26\}$$

(Note the similarity of this result to those derived in one dimension in Chapter 3.)

Since the previously-mentioned steady-state LQR is time-invariant, linear, and easily implementable, it is desireable to know whether the system in question will have a constant, positive-definite steady-state solution S(SS) as S(k) is iterated backward in time, for any choice of S(N). In short, Lewis (1986) states the following. Let the pair (A, √Q) be observable (i.e. let there exist some matrix L such that the poles of

$$A - \sqrt{Q}\,L$$

can be placed anywhere). Then it is necessary and sufficient that (A,B) be stabilizeable (i.e. a K exists such that

$$A - BK$$

has stable poles) for such a solution S(SS) to exist. Now let the pair (A, √Q) not be observable. Then the above relation between the stabilizeability of (A,B) and the existence of S(SS) may or may not hold. Stated another way, if (A, √Q) is not observable then we are not guaranteed anything.

When (A,B) is stabilizeable and when (A, √Q) is observable, the existence of a unique S(SS), and the resulting unique time-invariant feedback gain K leads to frequency domain analysis and design of the LQR. The Chang-Letov procedure accomplishes this for single-input systems with performance indices Q = qI and R = r, and with n x 1 transfer function matrix H(z). It can be shown that the "characteristic equation" G(z) of the overall closed-loop system, including the steady-state suboptimal feedback, is given by {4.27}:

$$G(z) = 1 + \frac{q}{r} H^T(z^{-1})\,H(z) \qquad \{4.27\}$$

The actual poles of the closed loop system are the *stable* roots of G(z) = 0. These roots can then be used for a pole placement controller design, if desired.

Figure 4.10 shows the "optimal root locus" for the system of parameters given by {4.20}. Note that for q in the range [-900, -400] the root of the system goes complex. But how can a one-dimensional system have complex roots? Which root do we choose? We shall address these questions in the following chapter.

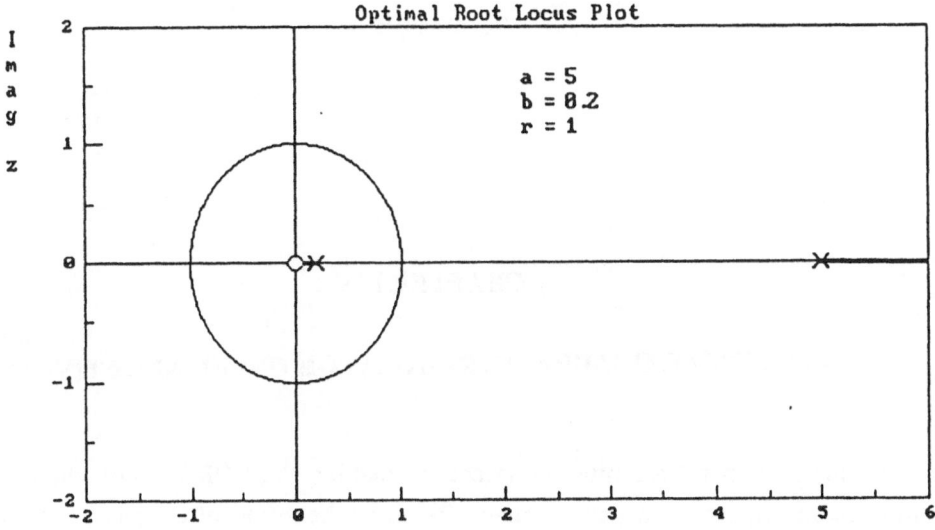

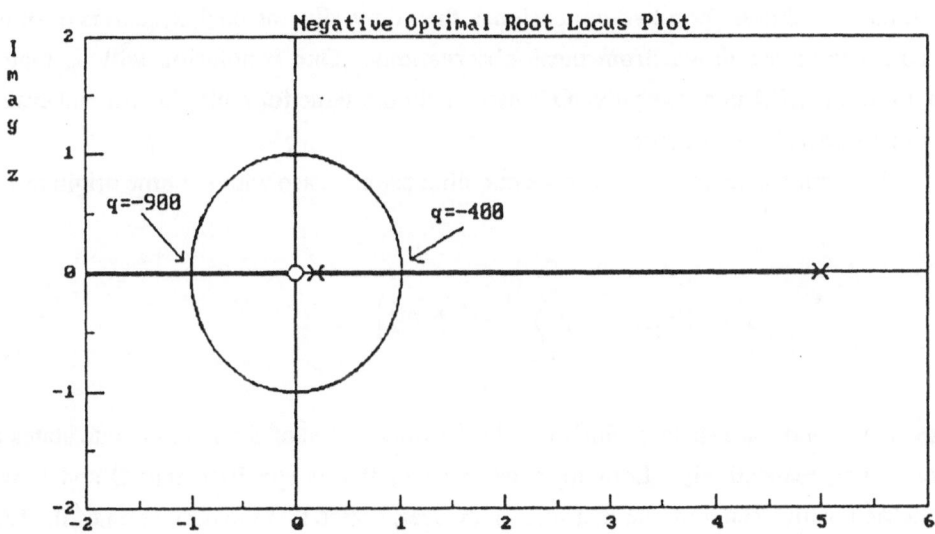

Figure 4.10

CHAPTER FIVE

THE LINEAR QUADRATIC REGULATOR UNDER NEGATIVE Q

In this section we examine the effects of making the LQR's intermediate-state weighting matrix, Q, take on negative values. Based on the results of Chapters 2, 3 and 4, the one-dimensional case for negative q is studied in detail. After the one-dimensional case is resolved, we shall extend our analysis to the multidimensional case in an empirical manner: a typical numerical case for a two-dimensional, then three-dimensional, system is examined and conclusions are drawn from these observations. One conclusion will be that the behavior of the LQR under negative Q is essentially the same for multidimensional systems as for one-dimensional systems.

Rewriting equation {4.1} for the one-dimensional case with the time origin at zero, i.e.

$$J = .5 \, s_N x_N^2 \; + \; .5 \sum_{k=0}^{N-1} \left[q x_k^2 + r u_k^2 \right] \tag{5.1}$$

we see that q and r are scalars which weight the importance of the intermediate states and system input, respectively. Looking back at {4.1} it was specified that Q and R were symmetric positive semidefinite matrices. However, notice that the derivation of the LQR, designed to minimize {4.1}, does not really require anything but symmetry (in order to take the partial derivatives of J with respect to x(k) and u(k)) of the matrices Q and R. Clearly, the restrictions on Q and R arose from practical considerations, since negative weighting of the intermediate states and input is of questionable physical meaning.

Letting r stay positive, we will consider the case with q negative. What does a negative q mean? One answer might be that it is for "evasion": staying away from a specified point (i.e. maximum squared error) at intermediate times but with a minimum of input energy and final state error. What effect does a negative q have on the system trajectory, and for what values does different behavior occur?

To answer these questions, we cast equation {4.15} into the one-dimensional form and mold it into the form of {2.1}

$$s_k = q + \frac{a^2 s_{k+1}}{1 + \frac{b^2}{r} s_{k+1}} \qquad \{5.2\}$$

or

$$s_k = \frac{(a^2 r + b^2 q) s_{k+1} + qr}{b^2 s_{k+1} + r} \qquad \{5.3\}$$

Clearly, this is of the form

$$s_k = \frac{a' s_{k+1} + b'}{c' s_{k+1} + d'} \qquad \{5.4\}$$

for which we have already derived an exact intricate closed-form solution and analysis in Chapter 3. Al that is left to do is substitute and observe. Let

$$\begin{aligned}
a' &= a^2 r + b^2 q \\
b' &= qr \\
c' &= b^2 \\
d' &= r
\end{aligned} \qquad \{5.5\}$$

As derived in Chapter 3, the equation {5.4} can have an oscillatory or a fixed point trajectory as dictated by the inequality

$$(a' - d')^2 \; <> \; -4b'c' \qquad \{5.6\}$$

If "less than" the trajectory is oscillatory and if "greater than" it is fixed-point. Substituting {5.5} into {5.6} for the oscillatory case, we obtain

$$(a^2 r + b^2 q - r)^2 < -4qrb^2 \qquad \{5.7\}$$

which becomes

$$b^4 q^2 + 2[a^2 + 1] r b^2 q + [a^2 - 1]^2 r^2 < 0 \qquad \{5.8\}$$

Note that {5.8}, an equation in q for a concave-upward parabola, is shown in Figure 5.1. The shaded region of Figure 5.1 represents the values of q which yield oscillatory behavior. Solving {5.8} we obtain {5.9}:

$$q_0, q_1 = -r \frac{(a + 1)^2}{b^2} \qquad \{5.9\}$$

It is interesting to note two facts yielded by {5.9}. First, the system will not be oscillatory whenever q and r have the same sign. Given that r is positive, this means that "strange" behavior will never occur for positive q. This corresponds to the theorem given in Chapter 4 concerning the existence of s(SS). Second, there is a mathematically obvious (from {5.9}) but intuitively mysterious region to the left of q_0 (i.e. more negative) for which the trajectory is fixed-point!

We investigate and verify {5.9} by a computer implementation of the system shown in Figure 5.2. From {5.9} we predict that the system will be oscillatory in the region

$$q = [-900, -400] \qquad \{5.10\}$$

Note how this corresponds to the values on the optimal root locus of this system in Figure 4.10.

Figures 5.3 through 5.11 show the trajectory of (a) the system output x(k), and (b) the system cost kernel matrix s(k) for several pertinent values of q. These sample plots yield the primary results of this book, and serve to verify our analysis and to reveal the nature of the system under these "strange" circumstances. Let us examine them closely.

Figure 5.3 shows a normal system at q = 10. Figure 5.4 shows a case for q = -395, still in the fixed-point region. Note that s(k) reaches a nice positive steady-state value and that the state trajectory is still optimal. Figure 5.5 shows a more degraded, but still "normal" system at q = -399, just before the oscillatory region.

At q = -405, Figure 5.6 shows that the system has become oscillatory. Note the "chaotic"-looking trajectory of s, the dynamics of which were fully explained in Chapter 3. The dynamics of x(k), the state trajectory, are described by the difference equation

$$x_{k+1} = \frac{k11 \sin k\beta + k12 \cos k\beta}{k21 \sin k\beta + k22 \cos k\beta} x_k \qquad \{5.11\}$$

where k11, k12, k21, and k22 are functions of a, b, q, and r, and where ß is the angle in radians of the (complex) eigenvalues of the matrix

$$
\begin{bmatrix} a^2r + b^2q & qr \\ b^2 & r \end{bmatrix}
$$
{5.12}

Figures 5.7 through 5.11 show the evolution of the systems as q is further decreased past the lower oscillatory limit q = -900. Some interesting regularities are shown in the oscillatory region of the system in Figures 5.6 and 5.7. For example, notice a type of "symmetry" around the oscillatory limits of q, which are q = -400 to q = -900: the plot of x(k) for q = -895 is apparently a higher frequency oscillation modulated by the plot of x(k) for q = -405. The practical importance of this region is questionable, however.

Figure 5.9 shows s(k) and x(k) when q has moved into the mysterious fixed-point region, q = -901. Two important points to note are that the cost kernel has a negative steady-state value (which again raises questions of physical relevance) and that there is also a symmetry present: x(k) for q = -901 is apparently a high-frequency oscillation modulated by the trajectory of x(k) for q = -399. Figure 5.10 also reveals this symmetry for the values of q being -395 and -905.

Finally, Figures 5.9 through 5.11 point out something that goes against our intuition. As we make q more negative (intuitively meaning we want to *raise* the squared error) we actually find upon inspection that our squared error decreases, and zero is reached more quickly, as if we had made a positive q more positive. Although this agrees with our symmetry observations, it raises doubts as to whether we are still *minimizing* the cost function in the LQR derivation when q is made negative. Figure 5.12 shows a summary of system behavior versus the parameter q.

Conclusions and regularities regarding the data given up to this point are now presented. Special attention is given to the practically important case of suboptimal steady-state feedback and its associated root-locus design and analysis techniques. These conclusions will then be verified and further illustrated by two and three-dimensional LQR systems.

For positive q, a steady-state value s(SS) is always obtained. This corresponds to normal operation. In Chapter 2 it would correspond to the map of s(k+1) vs. s(k) crossing the 45-degree line at some positive value, with a slope of magnitude less than unity. The optimal root locus technique is guaranteed to work, since q exists, (a, $\sqrt{q}$) is observable, and (a,b) is stabilizeable. (For this and all systems to follow, (A,B) will be assumed stabilizeable

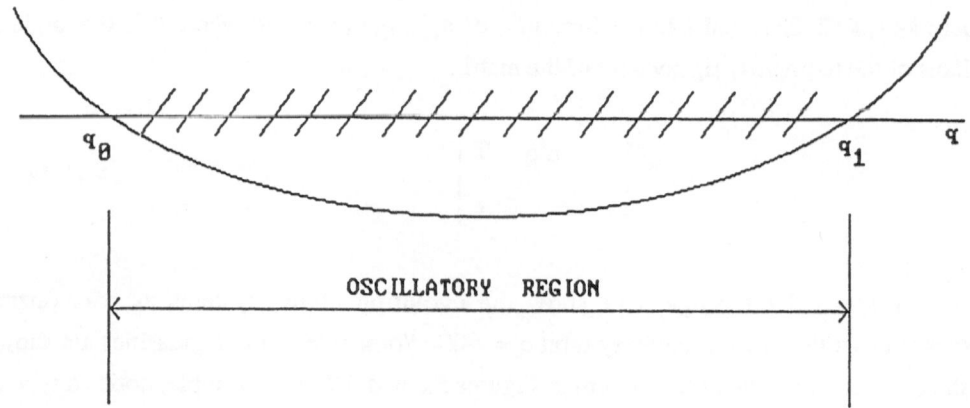

Figure 5.1

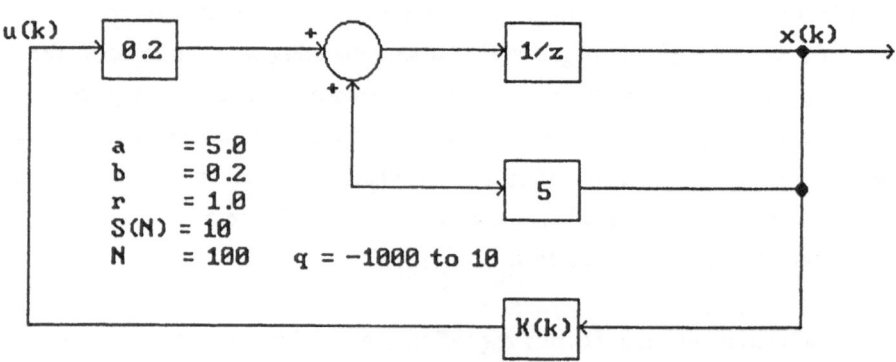

Figure 5.2

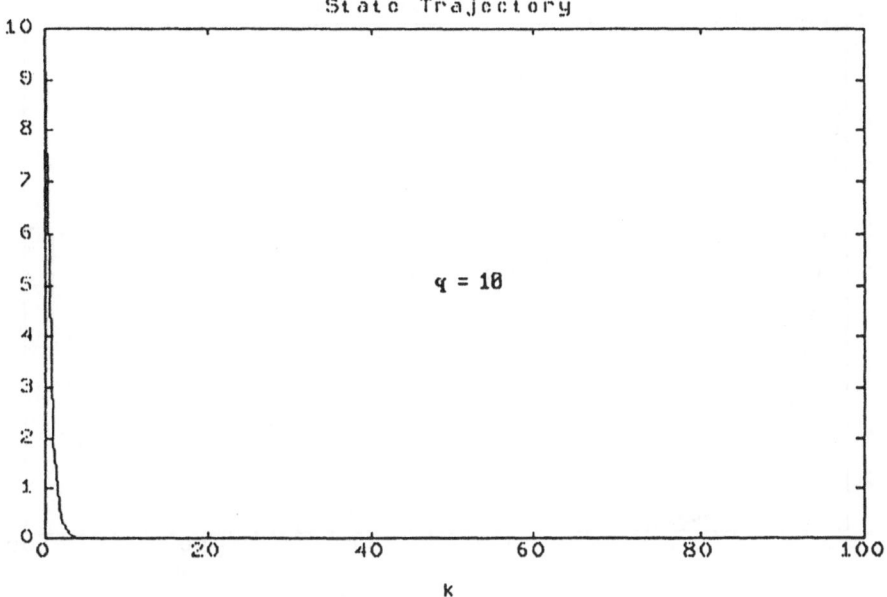

Figure 5.3a

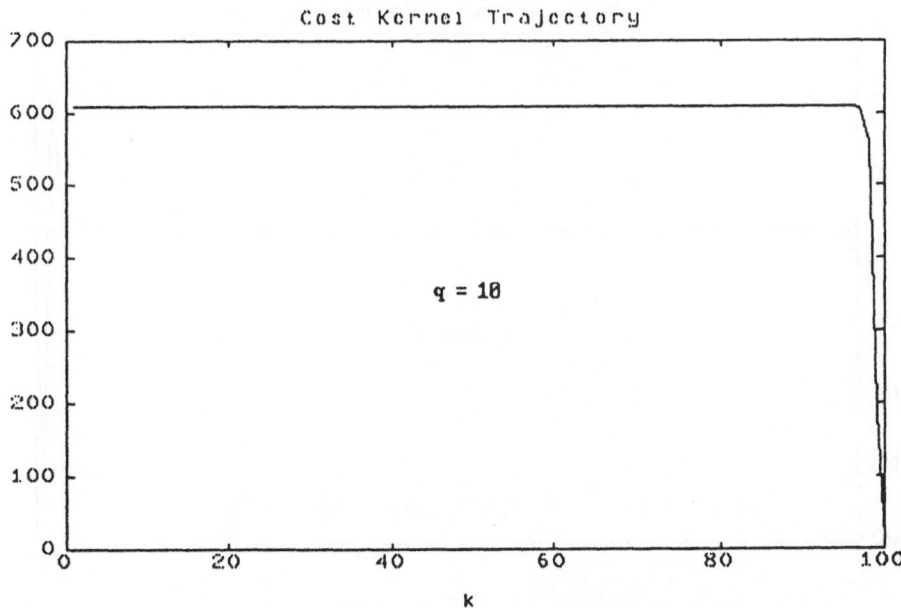

Figure 5.3b

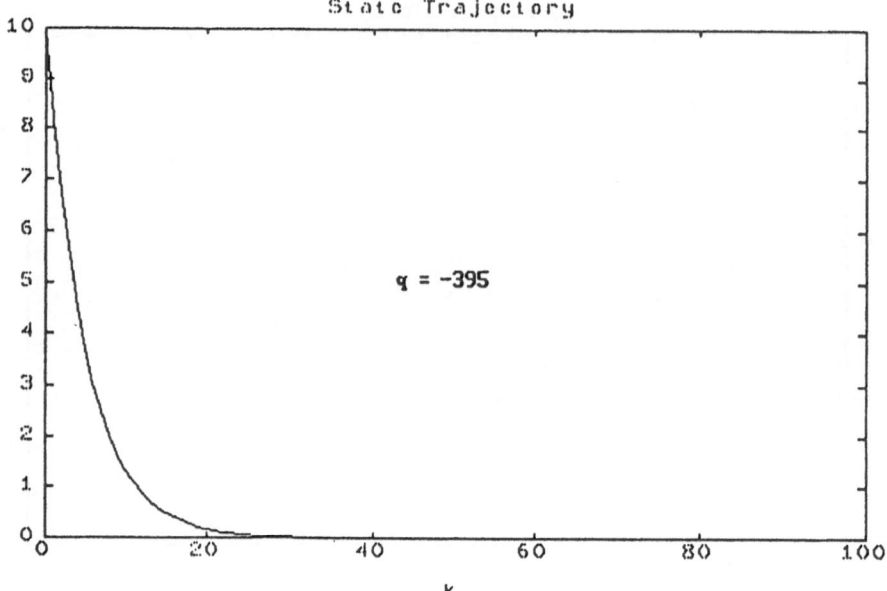

Figure 5.4a

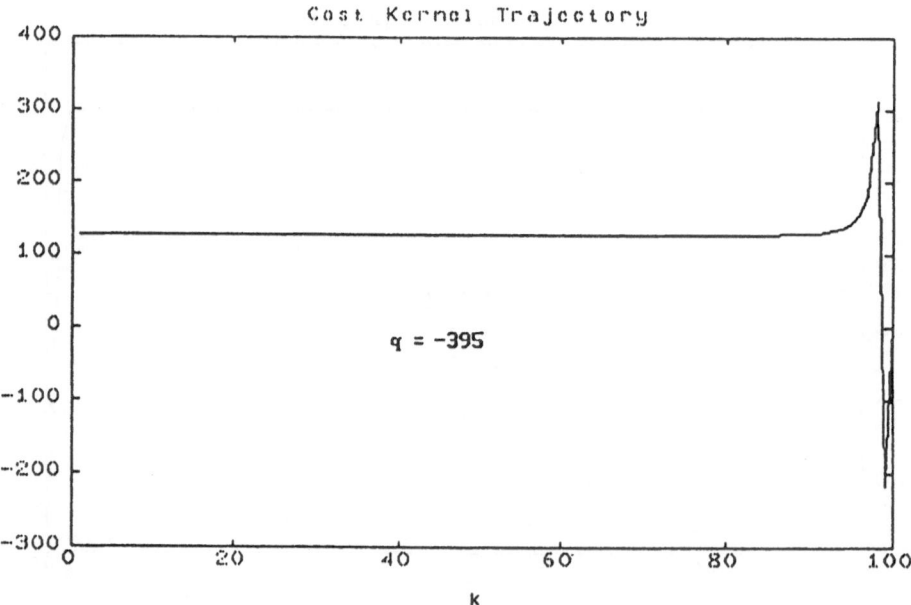

Figure 5.4b

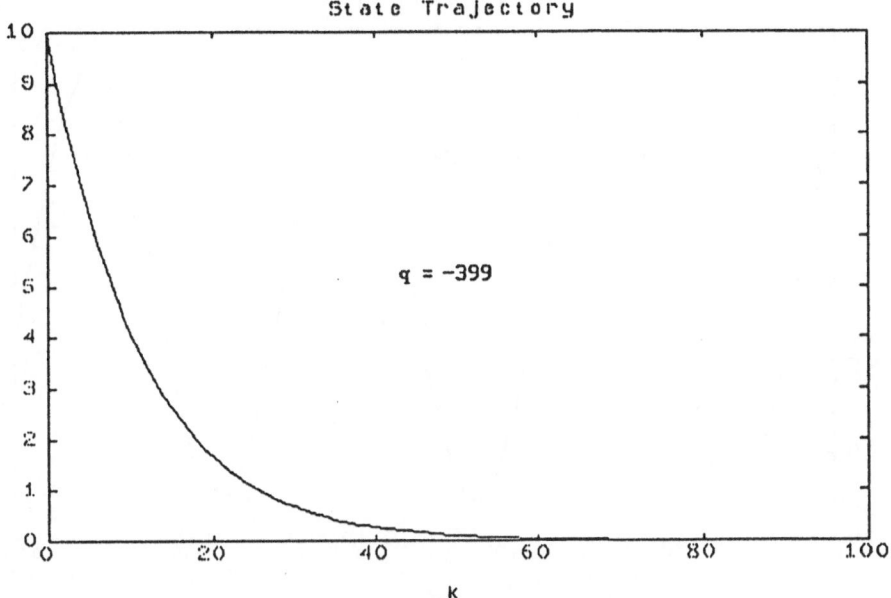

Figure 5.5a

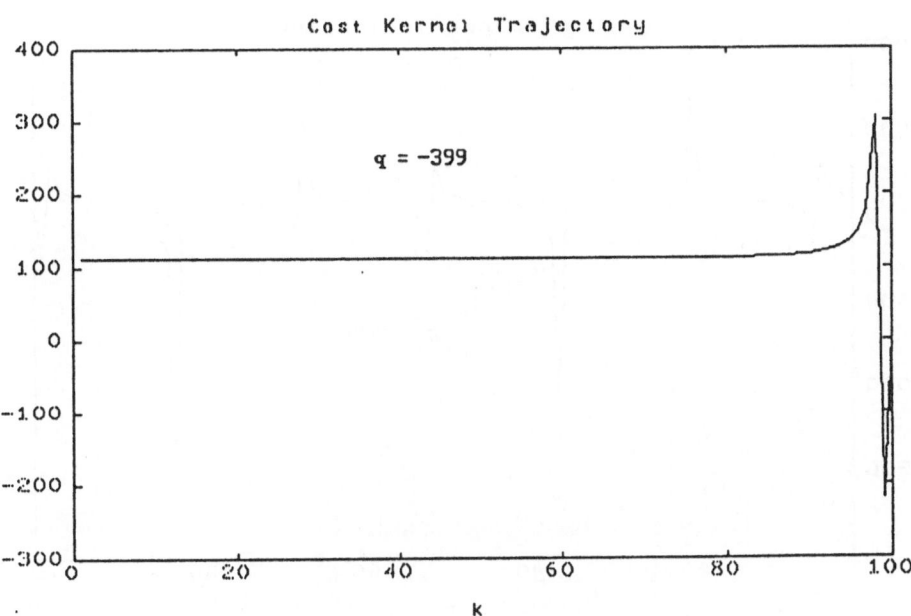

Figure 5.5b

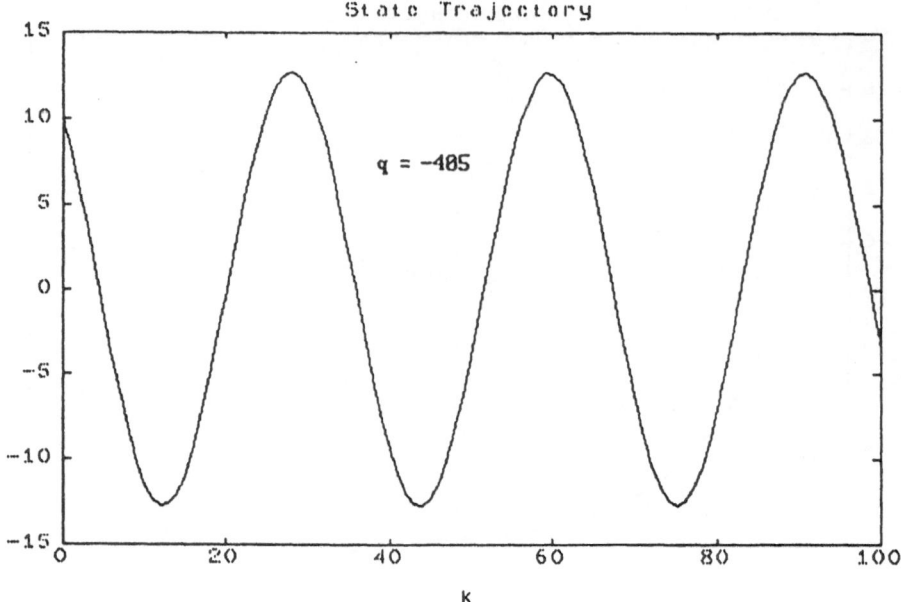

Figure 5.6a

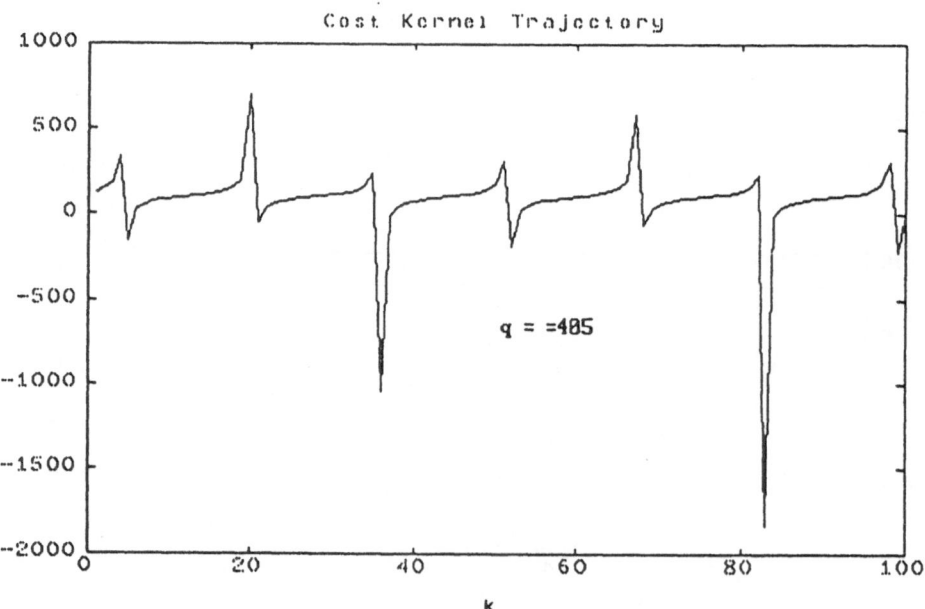

Figure 5.6b

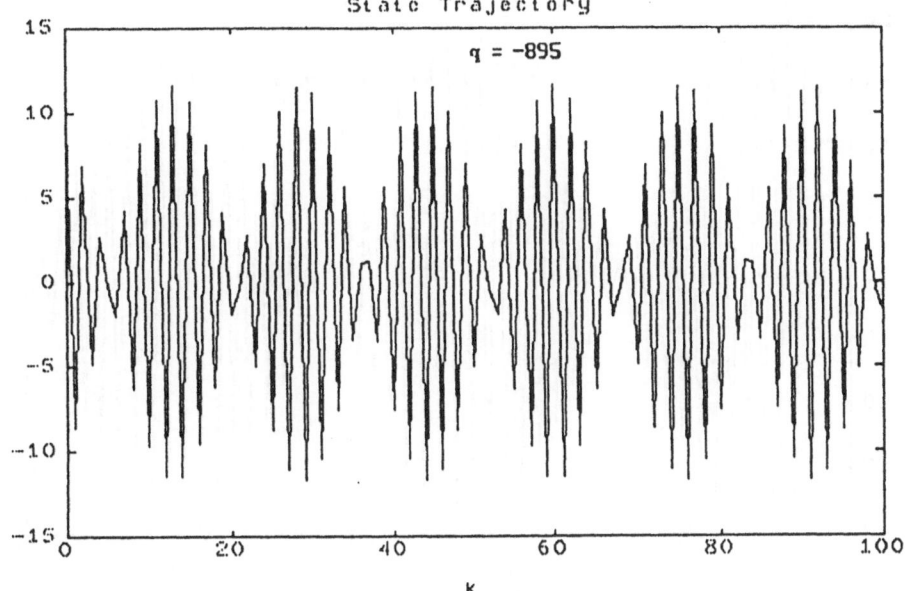

Figure 5.7a

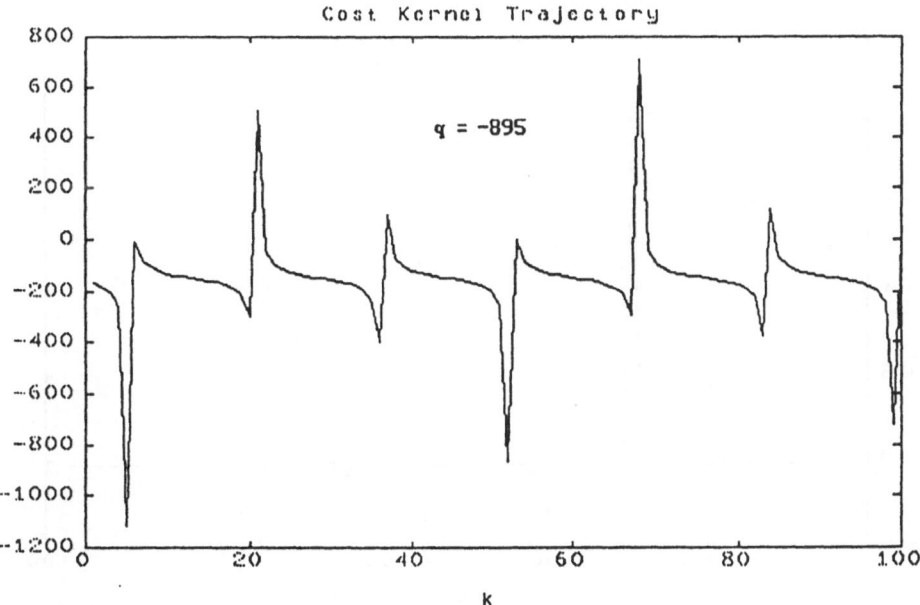

Figure 5.7b

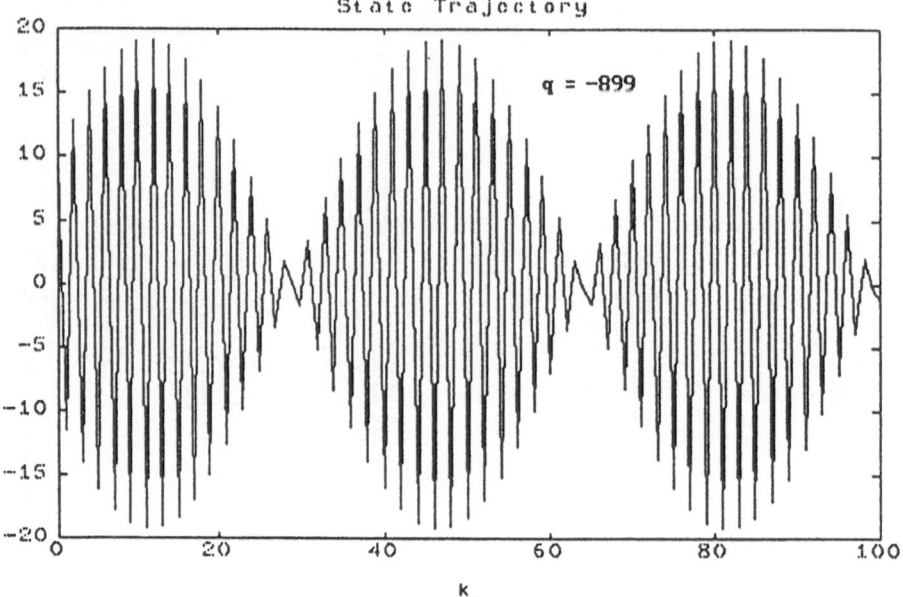

Figure 5.8a

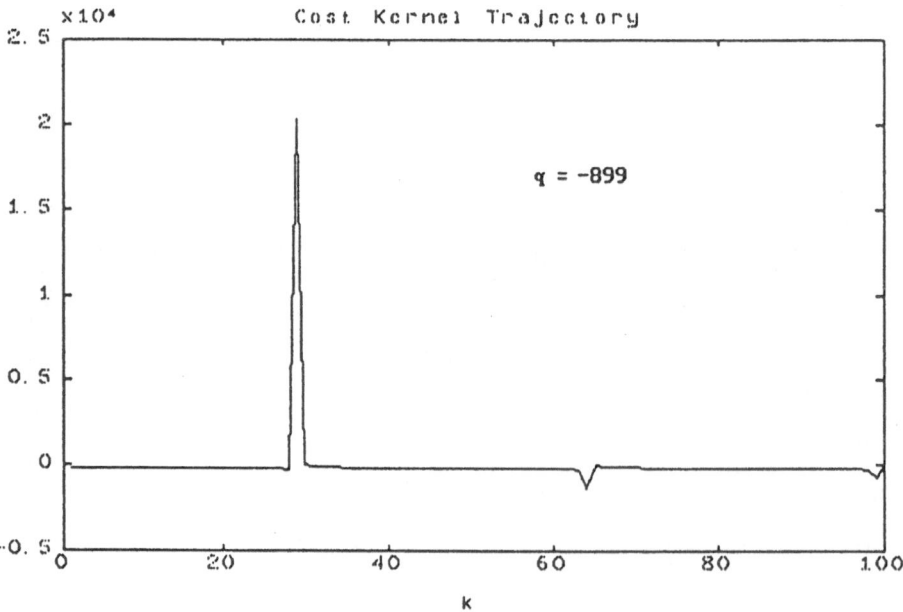

Figure 5.8b

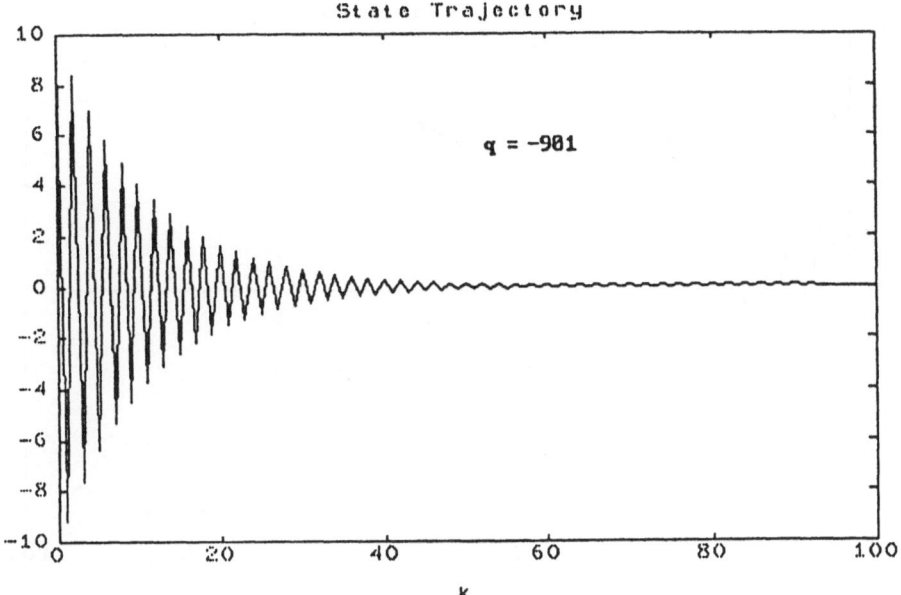

Figure 5.9a

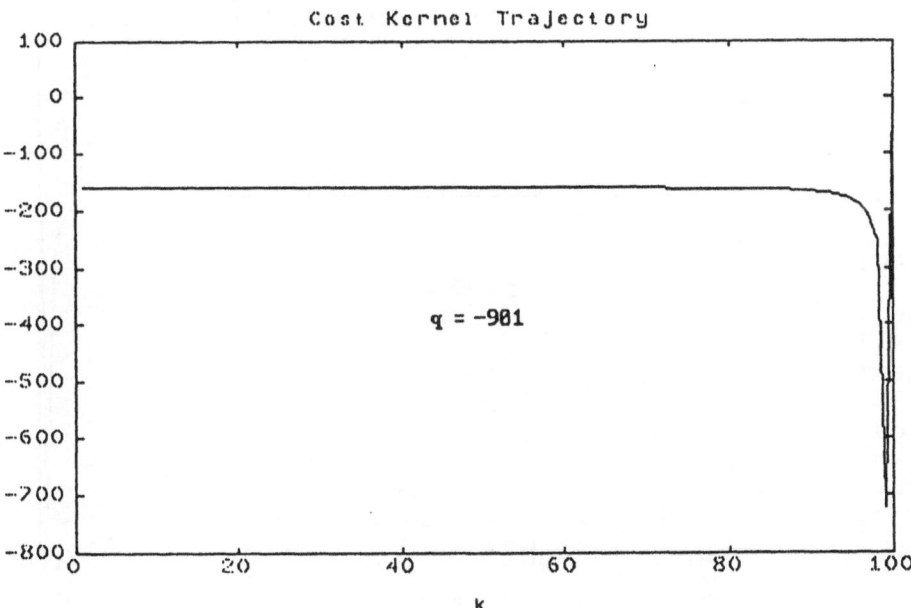

Figure 5.9b

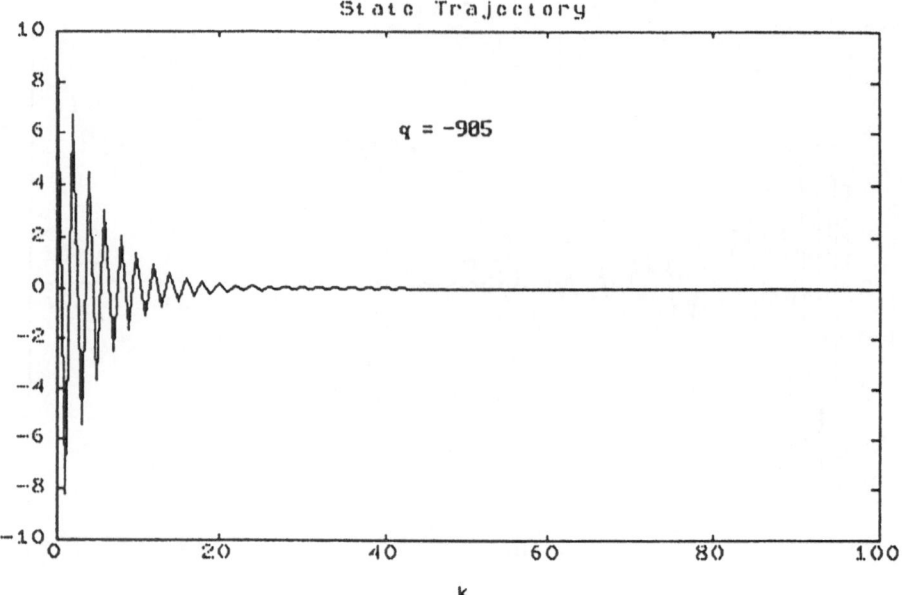

Figure 5.10a

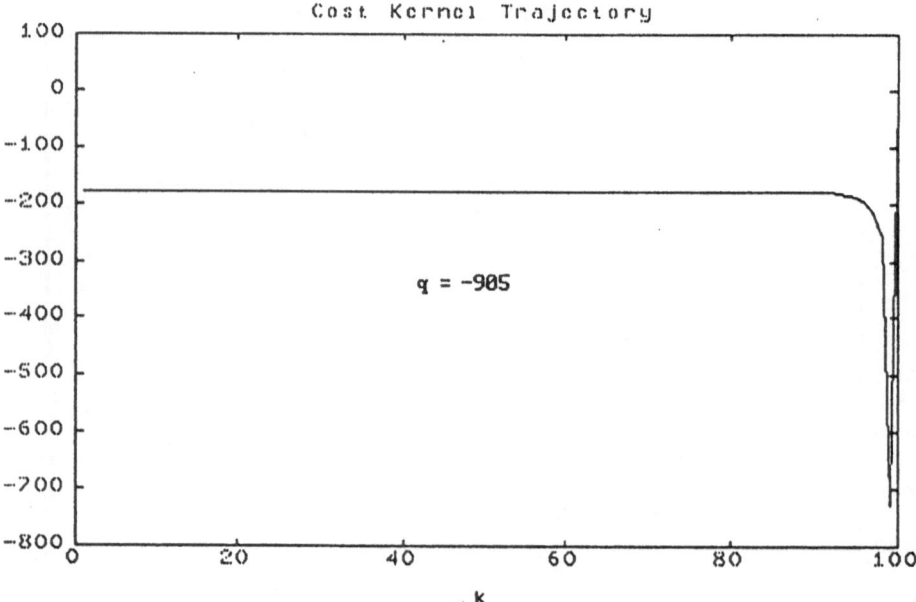

Figure 5.10b

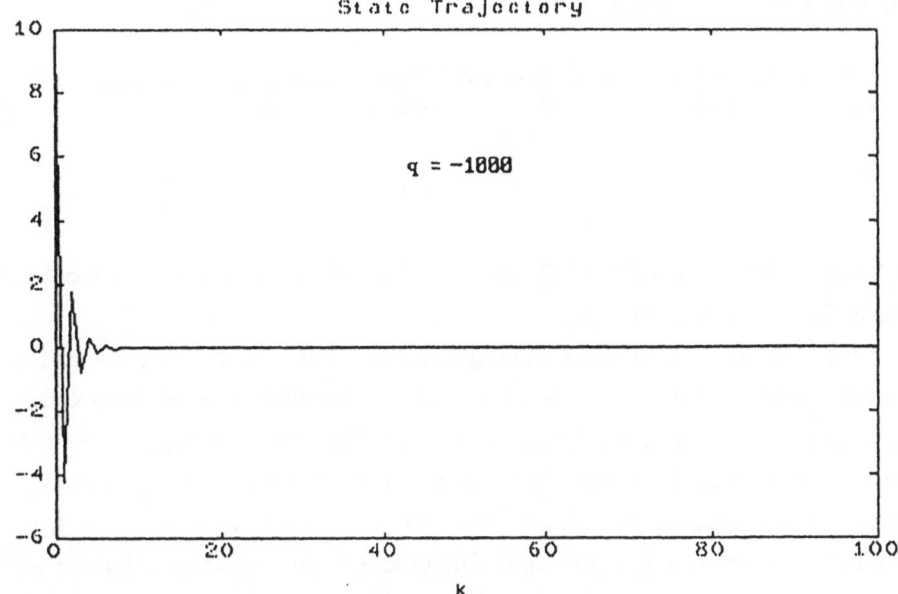

Figure 5.11a

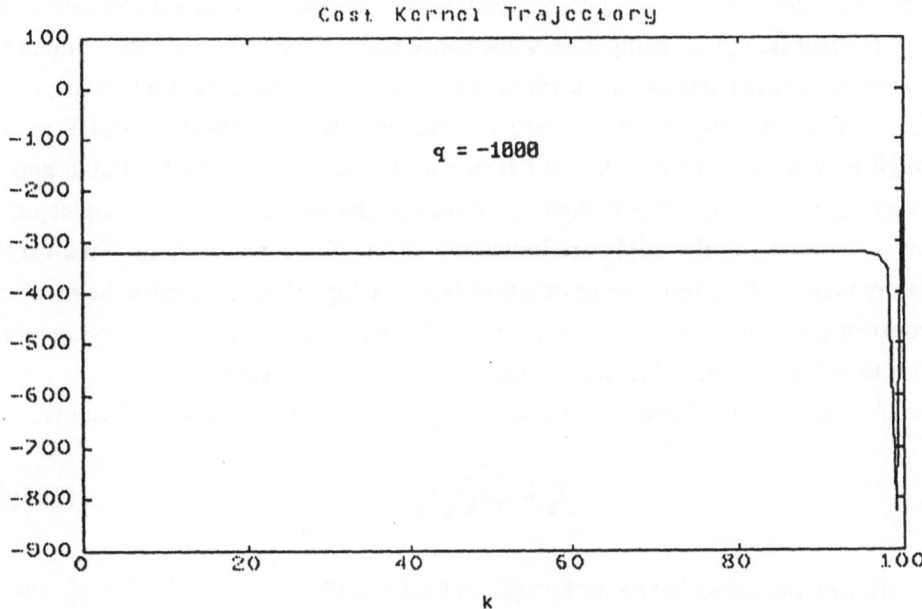

Figure 5.11b

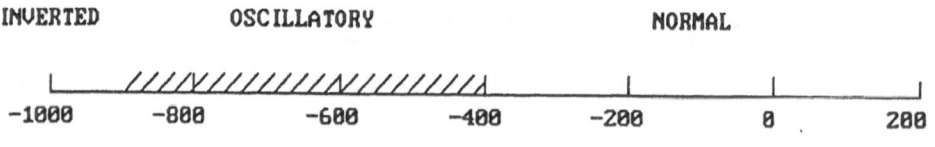

Figure 5.12

and (A,√Q) will be observable if √Q exists.) A steady-state, suboptimal feedback may always be found and used if desired.

Now when q goes negative, nothing really changes at first. The curve of s(k+1) vs. s(k) still crosses the 45-degree line in general at *some* positive value with slope magnitude less than unity, and the system trajectory is still quite "optimal" insofar as the cost function is still being minimized. However, since √q is not a real number, the pair (a, √q) is not observable in the usual sense. From Lewis' theorem discussed in Chapter 4, then, we conclude that we are no longer *guaranteed* that the optimal root locus technique will work anymore for a suboptimal system. Despite this, Figures 4.10, 5.2, and 5.3 still indicate "normal" operation for q being between 0 and -400. In short, the system "still works" but is operating in a theoretically unverifiable parameter region. And for the one-dimensional case, at least, we have analytically found the limits of this region.

As q sinks below -400, the trajectory of s(k) is no longer fixed-point. Its dynamics are oscillatory; the curve of s(k+1) vs. s(k) no longer intersects the 45-degree line anywhere.

The pair (a, √q) not being observable (since real √q does not exist), the optimal root locus technique, which depends on positive q for theoretically verifiable relevance, finally falters: nothing but complex "ghost" roots of some unrealizeable suboptimal LQR are left. Although they have intuitive relevance because (1) they are marginally stable and the time-varying system itself experiences sustained oscillations, and (2) their angles on the z-plane with respect to the origin are indicative of the "frequencies" of the time-varying oscillatory system, these roots are of questionable meaning. Why? Because they are roots of a constant-gain feedback system which cannot be built, because the time-varying system which is to be approximated does not aproach constant steady-state values.

As was shown in Chapter 4, the "cost-to-go" function is given by the formula:

$$J_k = \frac{1}{2}x_k^T s_k x_k \tag{5.13}$$

Noting the values taken on by s(k) and x(k) in Figures 3.6 and 3.7, how can the remaining cost at a later time be greater than the remaining cost at an earlier time? It cannot. Yet the plots indicate that this is the case, so equation {5.13} is apparently invalid for the case q = -400 to q = -900. But equation {5.13} was based on the principles of LQR operation. We must conclude, therefore, that these principles are violated when q is in the "strange" parameter region and, simply stated, our system is no longer a linear quadratic regulator: it no longer minimizes the overall cost function J of equation {4.1}.

At values below q = -900, however, the LQR system seems to work once again; in fact, the lower q is made, the better it works! Even though $\sqrt{q}$ does not exist, the optimal root locus technique now works once again, and looking at the locations of the roots in Figure 4.10 and the trajectories for q < -900, the roots are again stable. In Chapter 3, our s(k+1) vs. s(k) plot once more crosses the 45-degree line; however, it is the negative branch of the hyperbola that crosses it this time. A negative cost kernel matrix implies a negative remaining cost, as implied by {5.13}. That is, the cost-to-go is negative everywhere. Although physically questionable, this explains why the trajectory gets "better" as we continue to decrease q; it is explained from the fundamental definition of the LQR; and it leads to a central conclusion of this paper:

Simply put, the LQR is not designed to necessarily *minimize* the cost function

$$J = .5\, x_N^T S_N x_N + .5 \sum_{k=0}^{N-1} \left[x_k^T Q x_k + u_k^T R u_k \right] \qquad \{4.1\}$$

Rather, it is designed merely to set J's derivatives with respect to each of the quantities x(k), u(k), and the Lagrange multipliers λ(k) equal to zero, and it implements the recursion equations necessary to do this. The LQR procedure says nothing of *second* derivatives, and so it does not *necessarily* minimize J for all performance indices Q, R, and S(N).

In other words, the LQR serves only to keep the overall system operating at some "critical point", some "extremum", in a many-dimensional cost space. When Q and R are symmetric and positive semidefinite, then this critical point is our desired absolute minimum. Also, there exist some non-positive definite regions of Q (near the positive definite ones) for which the LQR still focuses in on this absolute minimum, and thus for which operation is still normal.

When Q is no longer positive definite *and* is in the oscillatory region, this critical point is still some extremum, but is now merely a local one; it is neither an absolute maximum or minimum in our multi-dimensional cost space. The operation of the LQR in this parameter region, then, is of questionable meaning, although some interesting behaviors result. We refer to this region of operation as the "oscillatory" region.

When Q goes far enough negative-definite, into our third region of operation, the "critical point" which the LQR locates is again an absolute extremum, but this time it is an absolute *maximum*. The Q matrix has "drowned out" the R matrix and causes the cost function J and, through a negative-definite S(k), the cost-to-go function J(k), to become negative. As Q is further decreased, the trasjectory x(k) is driven to closer to zero faster, *maximizing* the *negative* performance index value. We shall refer to this case as the "inverted" region of operation. Whether or not it is "optimal" is a question of definition. Under the formal definition of LQR operation (the minimization of a performance index consisting of the sum of weighted input energy, intermediate-state squared error, and final-state squared error) it is not.

Again, the key thought of this conclusion is that the LQR merely sets the cost function's derivatives equal to zero; whether it maximizes, minimizes, or does neither depends on the region of operation as dictated by Q and R.

Based on these ideas, we conclude that the case of negative Q is of questionable use, however interesting it is: the "intermittent" region is a violation of LQR principles, and the "inverted" region achieves nothing that cannot be done in the normal region. Thus, negative Q can probably not be used for "evasion" purposes because there is nothing optimal about the evasive regions, and there is nothing evasive about the optimal regions.

The above conclusions and observations also apply to multidimensional systems. No new behaviors arise with increasing dimension; although intermittent in its oscillatory region, the LQR is not chaotic, in which case new behaviors would arise. Presented next are experimental verifications for two- and three-dimensional systems.

The data is arranged as follows. First, the system diagram and parameters are given (i.e. Figure 5.13 (2-D case) & 5.23 (3-D case)). Q is chosen as a diagonal matrix, and one of the diagonal elements is changed as the others are held constant, taking the system through the "normal", "oscillatory", and "inverted" modes of operation. The corresponding cost kernel matrix S(k) and state vector trajectories are shown (i.e. Figures 5.14 through 5.20 & Figures 5.24 through 5.32). Following this, a map which shows the regions of LQR operations versus the diagonal parameters of Q is presented (i.e. Figure 5.21 & Figure 5.33). Finally, the optimal root locus plots of the system is shown (i.e. Figure 5.22 & 5.34); the overall system roots are those which are stable.

Caution is advised in comparing these optimal root locus plots to the other data, for the optimal root locus technique that was used restricted the values of the Q matrix to $Q = qI$, where q is an adjustable real number and where I is the identity matrix. These values represent subspaces on the parameter region plots along the lines $q11 = q22 = q33$. The consistency between the two types of data representations (optimal root locus plot and parameter-region plot) is evident by noting the values of q which represent "crossover" regions from normal to oscillatory and from oscillatory to inverted operation.

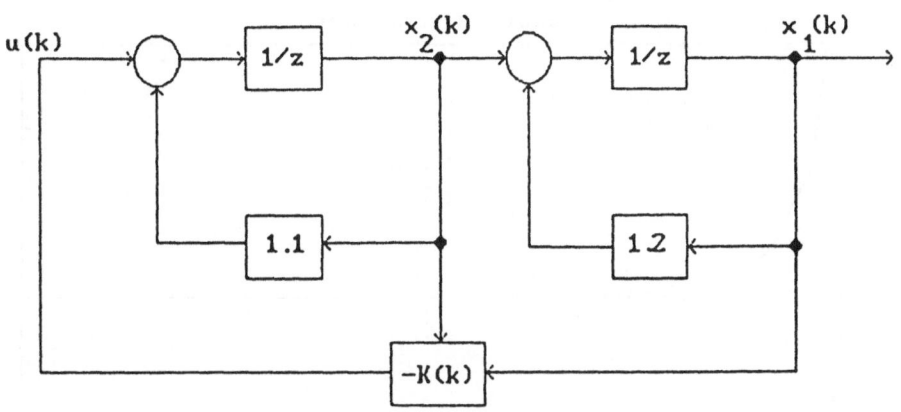

$$A = \begin{bmatrix} 1.2 & 1.0 \\ 0.0 & -1.1 \end{bmatrix} \qquad B = \begin{bmatrix} 0 \\ 1 \end{bmatrix}$$

$$Q = \begin{bmatrix} q_{11} & 0 \\ 0 & q_{22} \end{bmatrix} \qquad R = [5]$$

$$S(N) = \begin{bmatrix} 10 & 0 \\ 0 & 10 \end{bmatrix} \qquad N = 50$$

$$x(0) = \begin{bmatrix} x_1(0) \\ x_2(0) \end{bmatrix} = \begin{bmatrix} 5 \\ 0 \end{bmatrix}$$

Figure 5.13

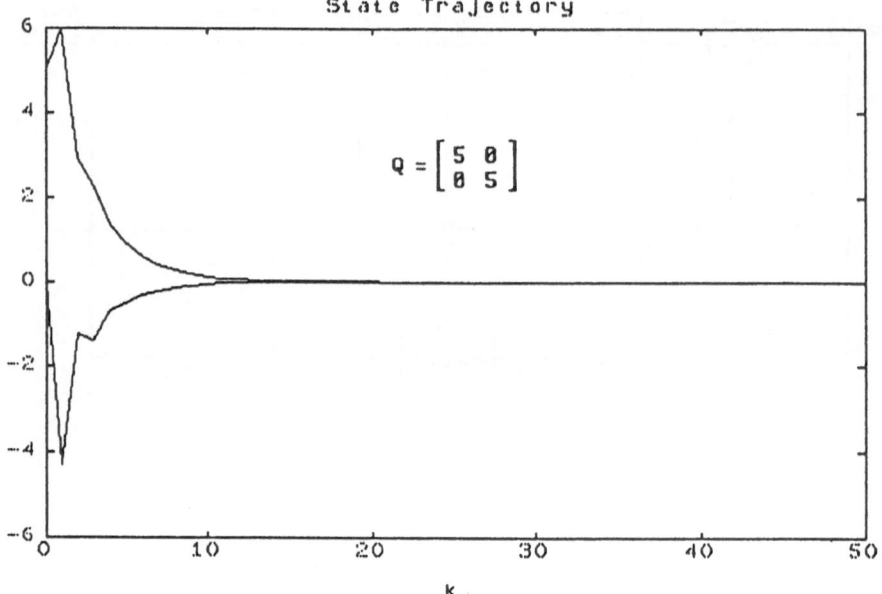

Figure 5.14a

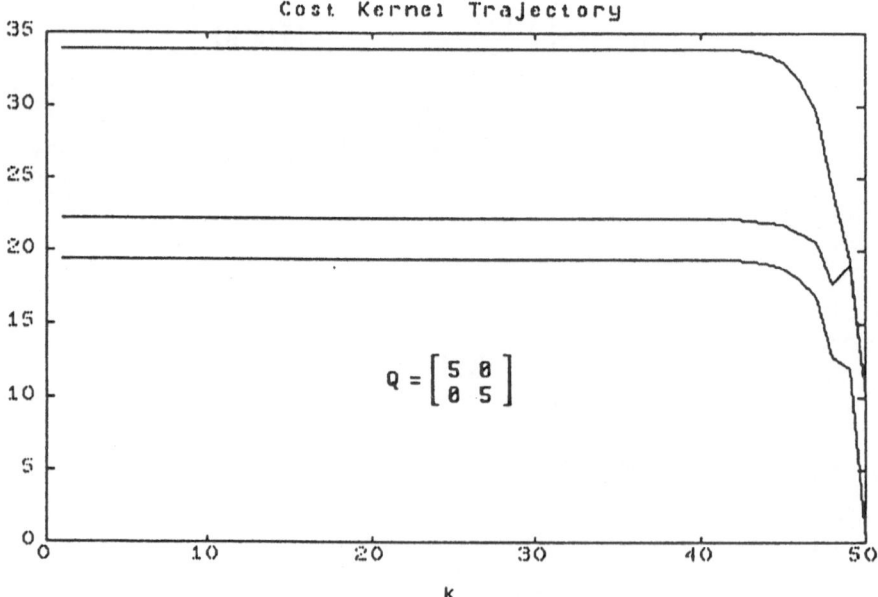

Figure 5.14b

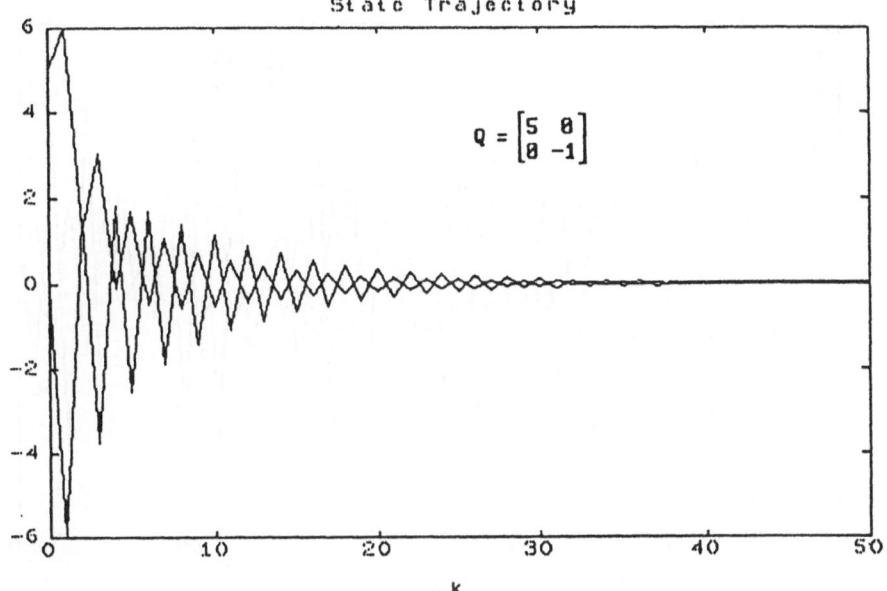

Figure 5.15a

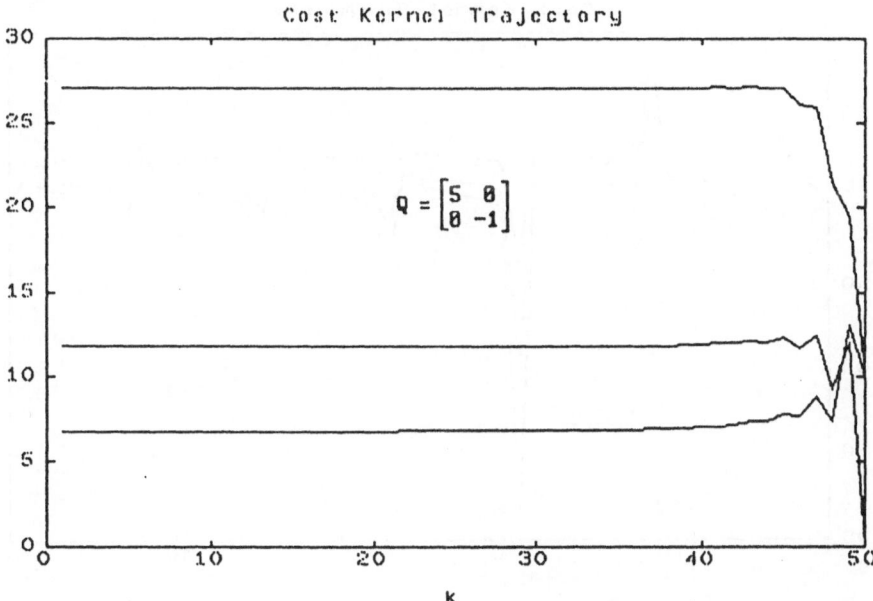

Figure 5.15b

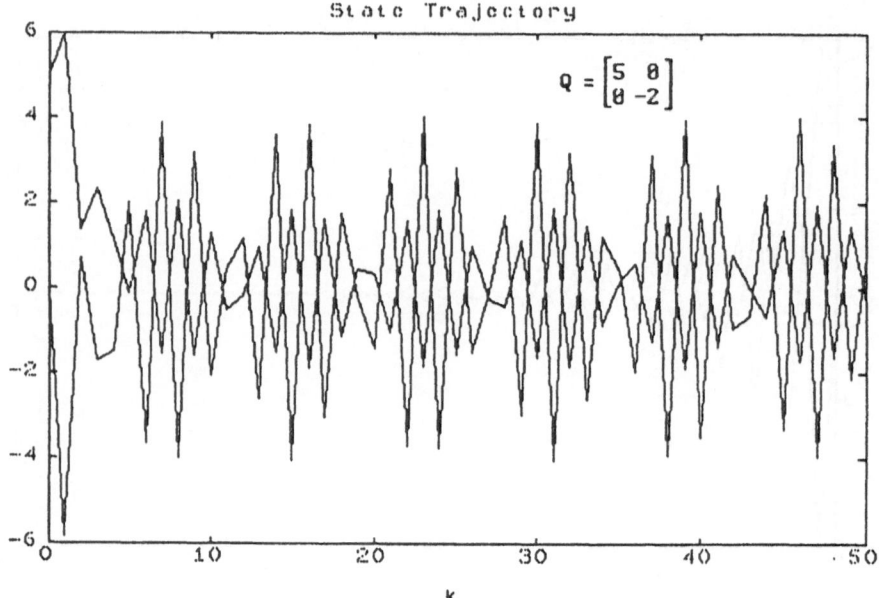

Figure 5.16a

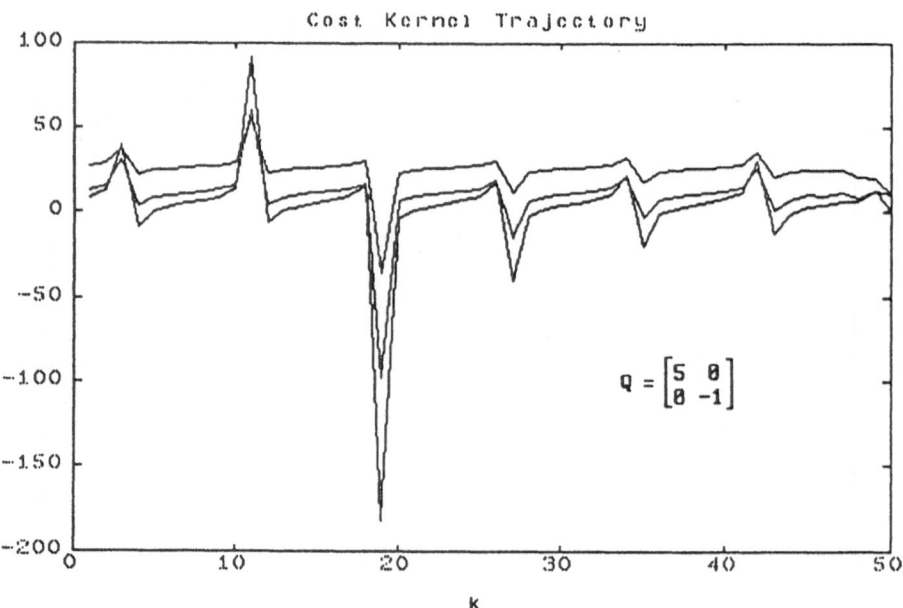

Figure 5.16b

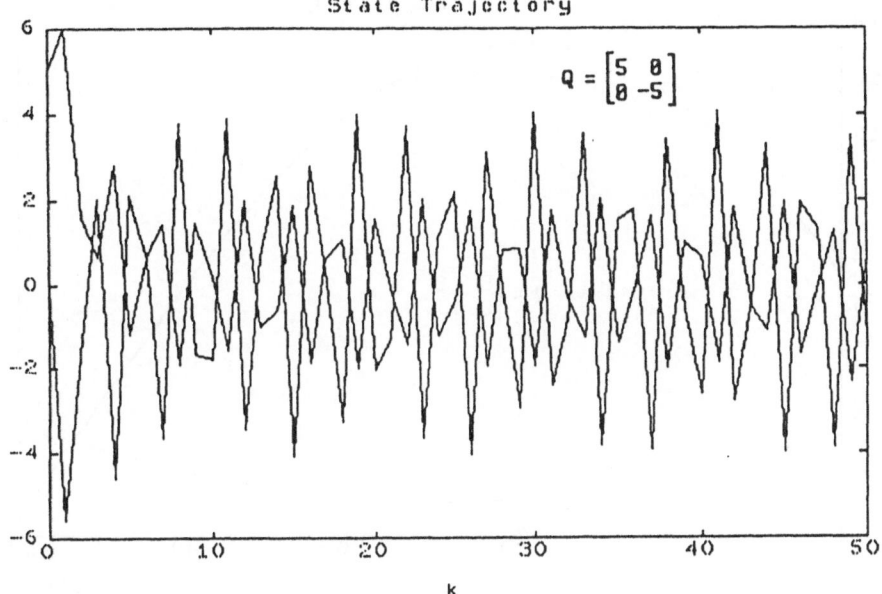

Figure 5.17a

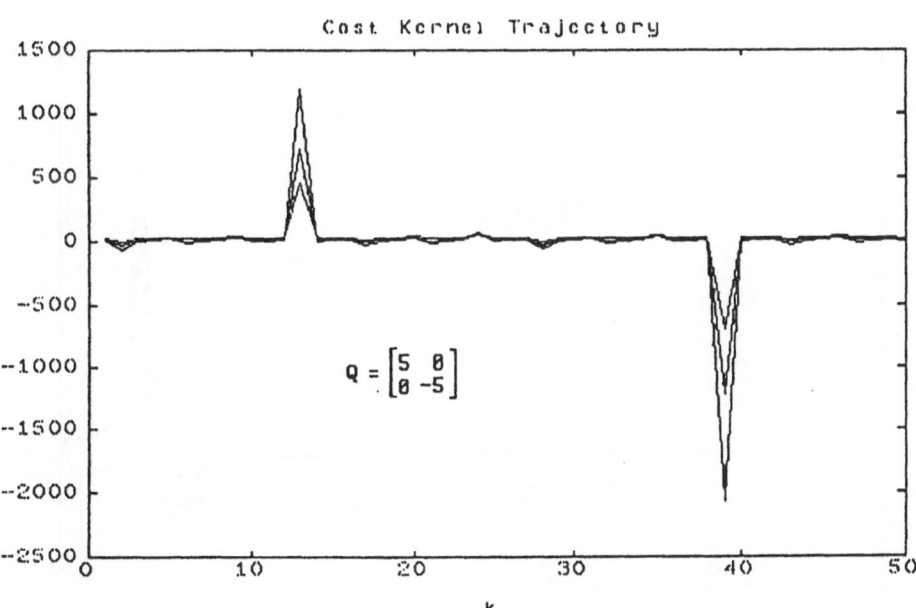

Figure 5.17b

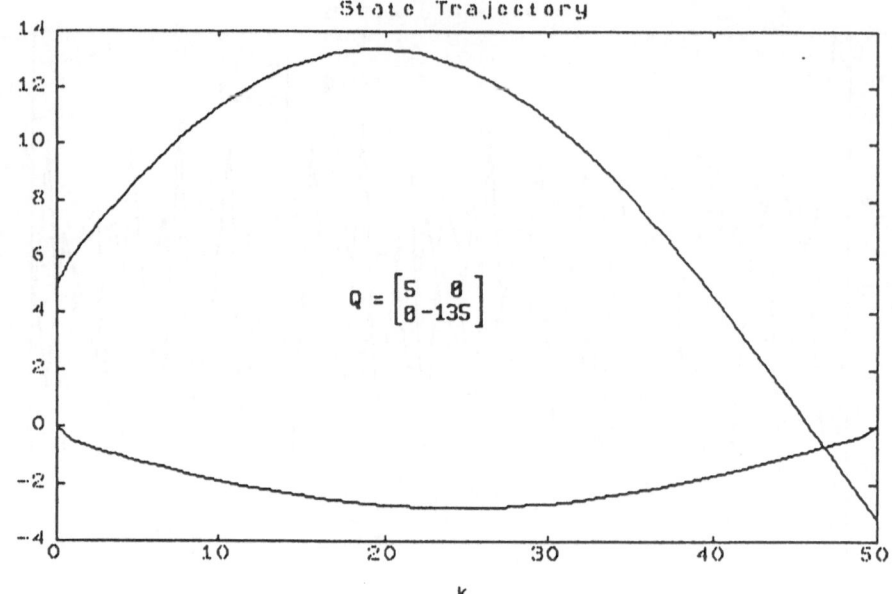

Figure 5.18a

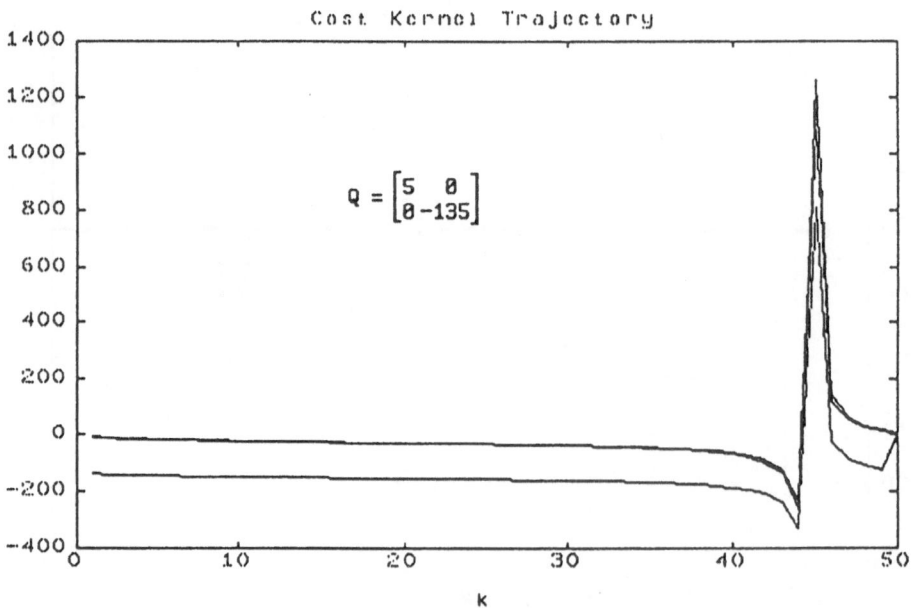

Figure 5.18b

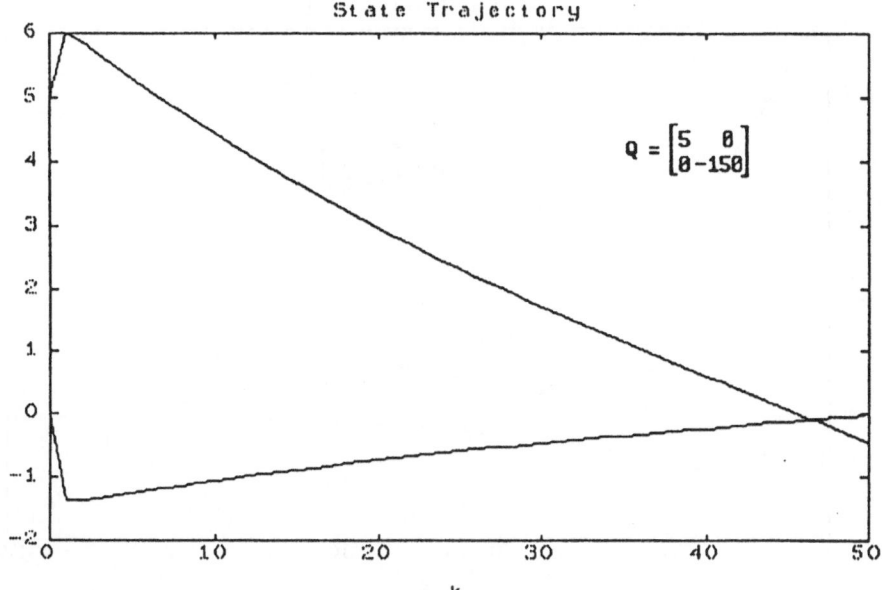

Figure 5.19a

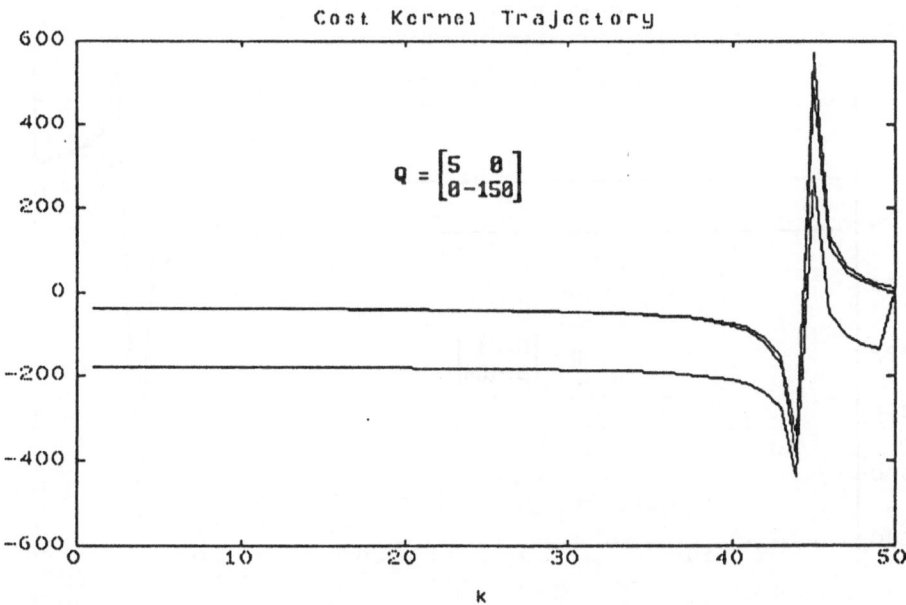

Figure 5.19b

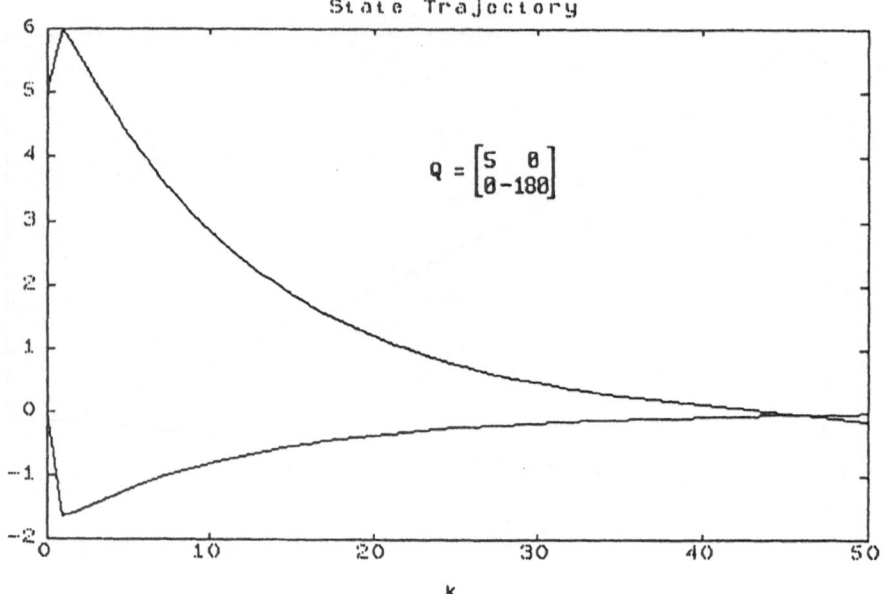

Figure 5.20a

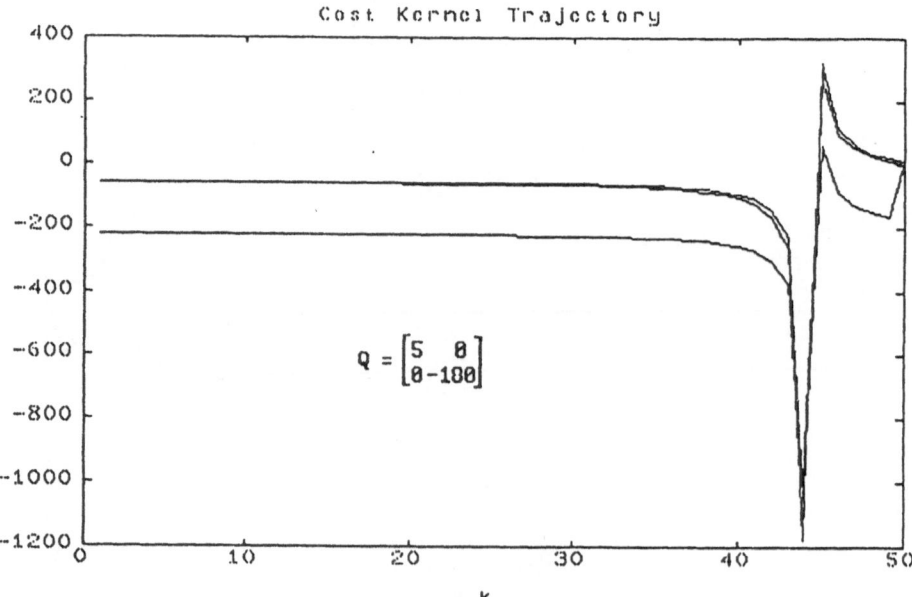

Figure 5.20b

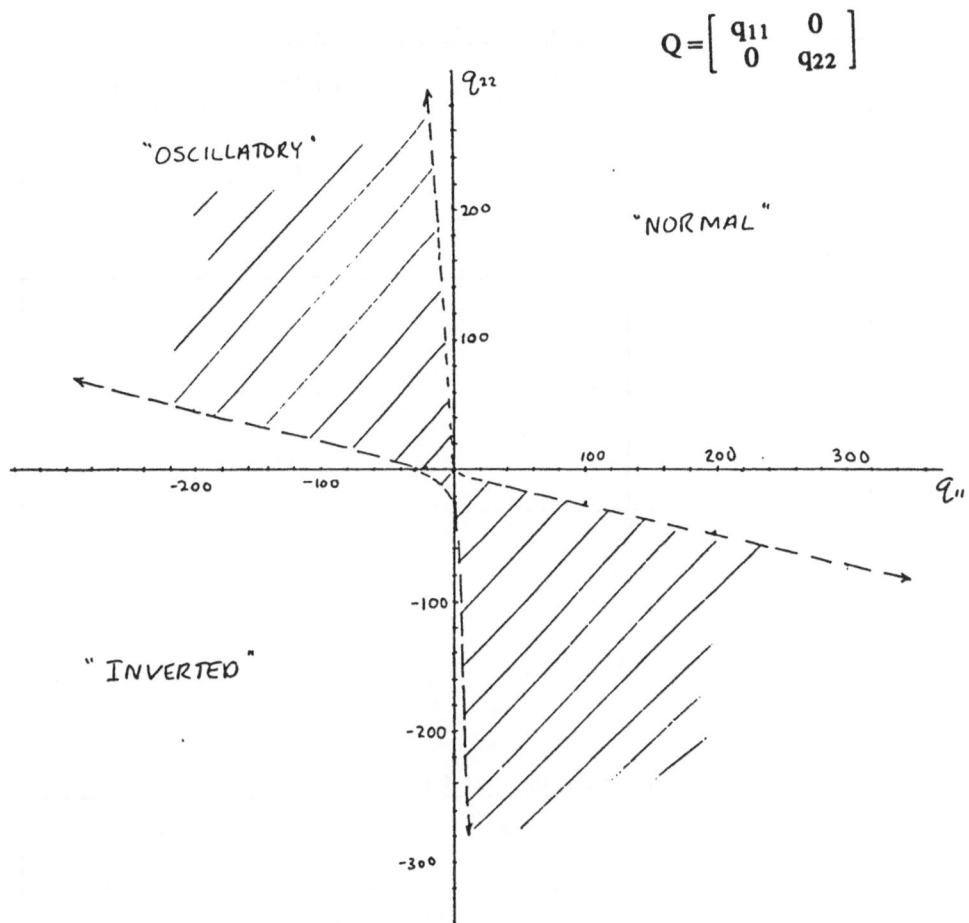

Figure 5.21 - Parameter region plot for the
2-dimensional system. Upper right region
is normal operation. Shaded region is
oscillatory. Lower left region   is inverted
LQR operation.

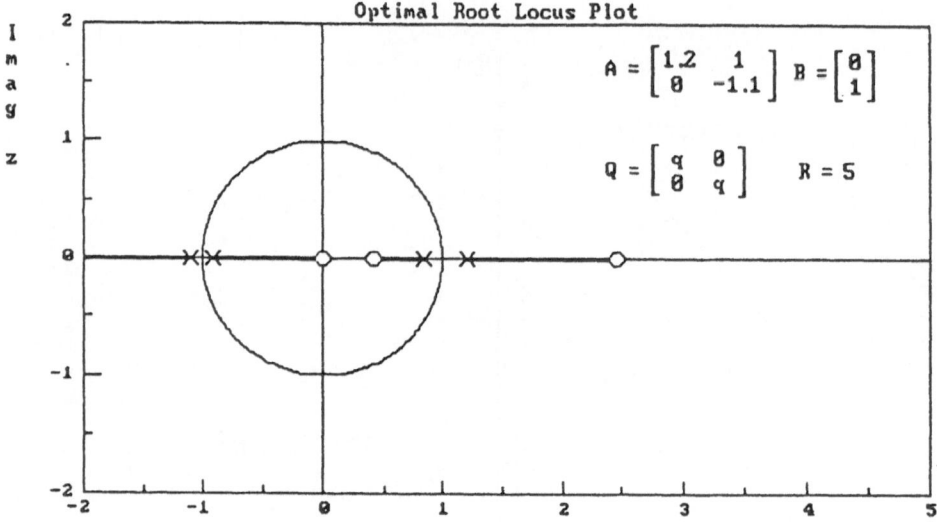

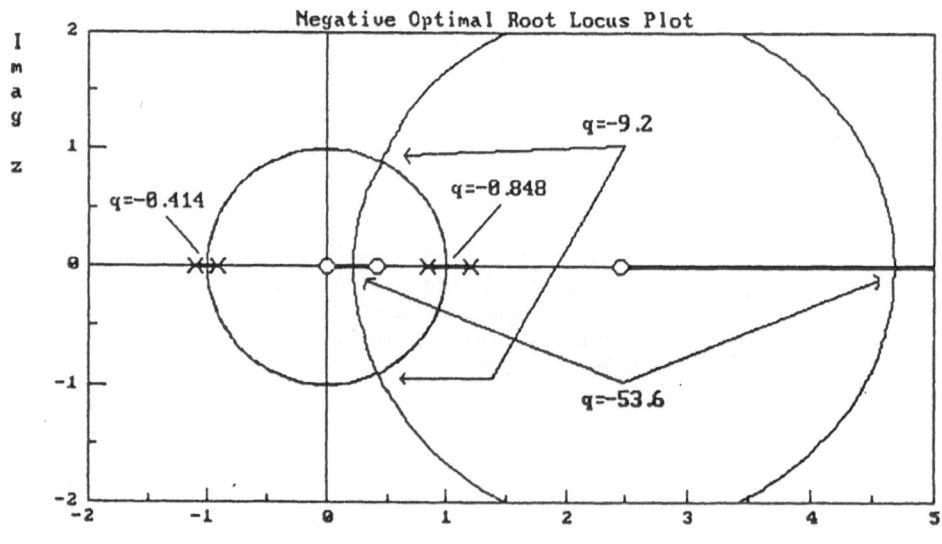

Figure 5.22

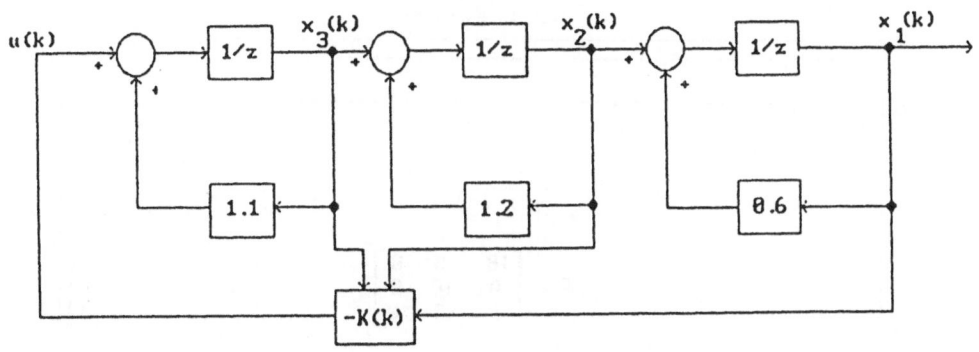

$$A = \begin{bmatrix} -0.6 & 1.0 & 0.0 \\ 0.0 & 1.2 & 1.0 \\ 0.0 & 0.0 & -1.1 \end{bmatrix} \qquad B = \begin{bmatrix} 0.0 \\ 0.0 \\ 1.0 \end{bmatrix}$$

$$Q = \begin{bmatrix} q_{11} & 0 & 0 \\ 0 & q_{22} & 0 \\ 0 & 0 & q_{33} \end{bmatrix} \qquad R = [5]$$

$$S(N) = \begin{bmatrix} 10 & 0 & 0 \\ 0 & 10 & 0 \\ 0 & 0 & 10 \end{bmatrix} \qquad N = 50$$

$$x(0) = \begin{bmatrix} 10 \\ 5 \\ 0 \end{bmatrix}$$

Figure 5.23

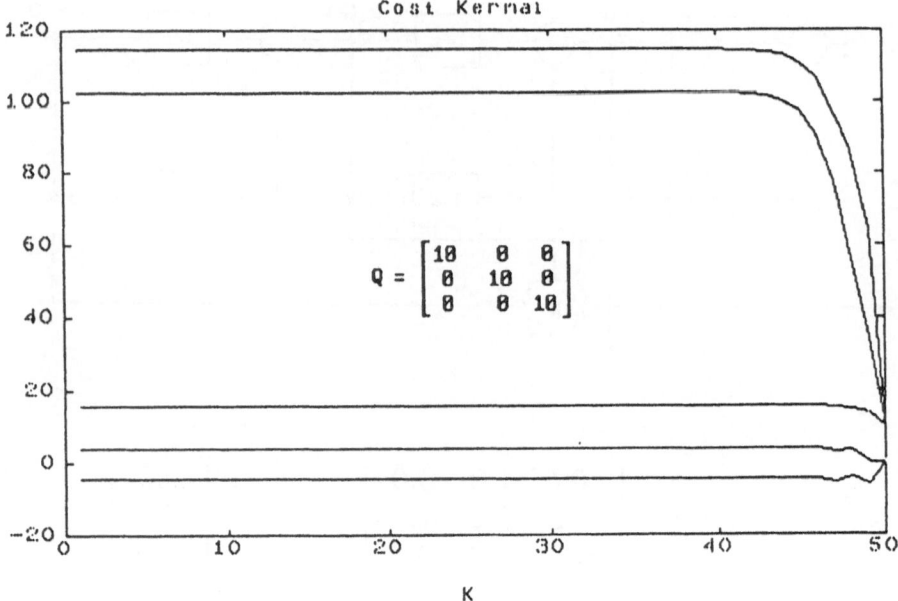

Figure 5.24a

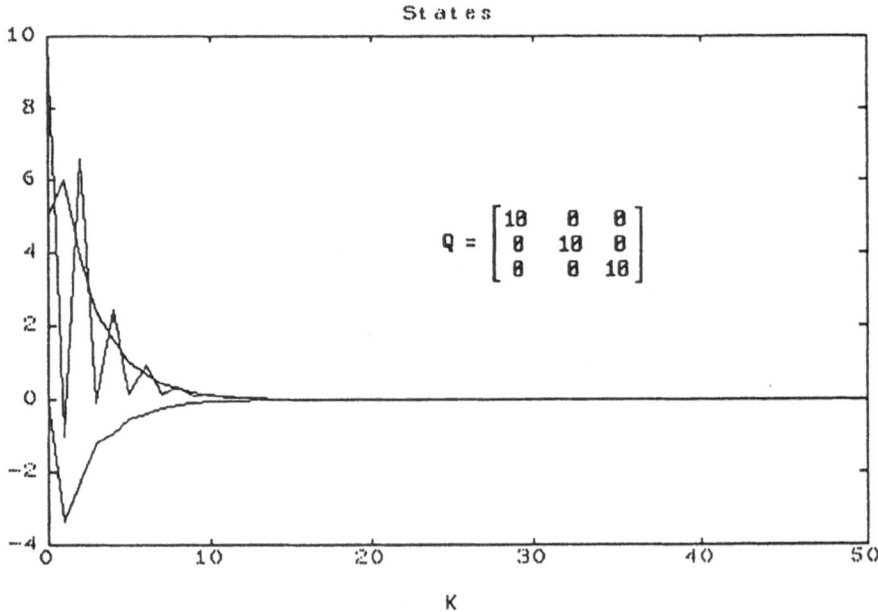

Figure 5.24b

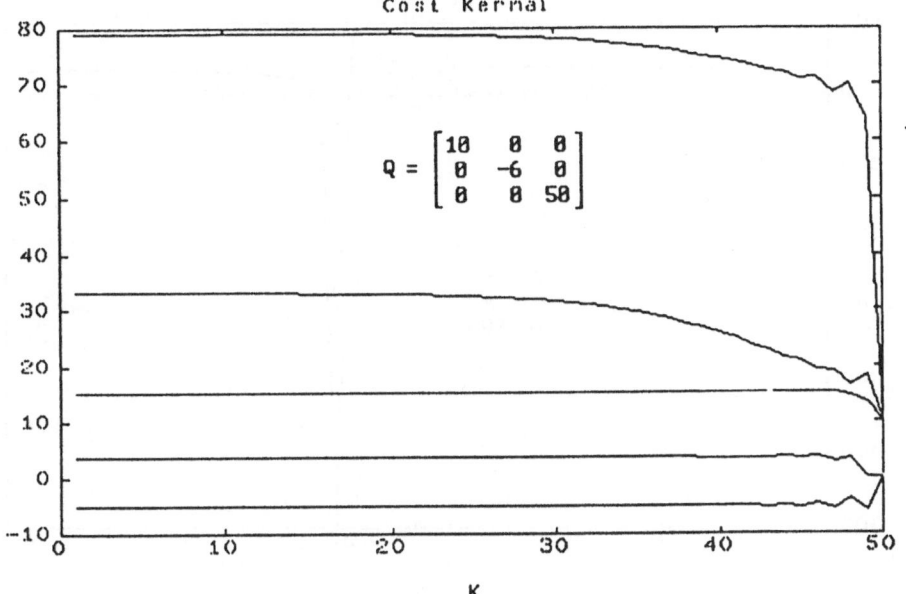

Figure 5.25a

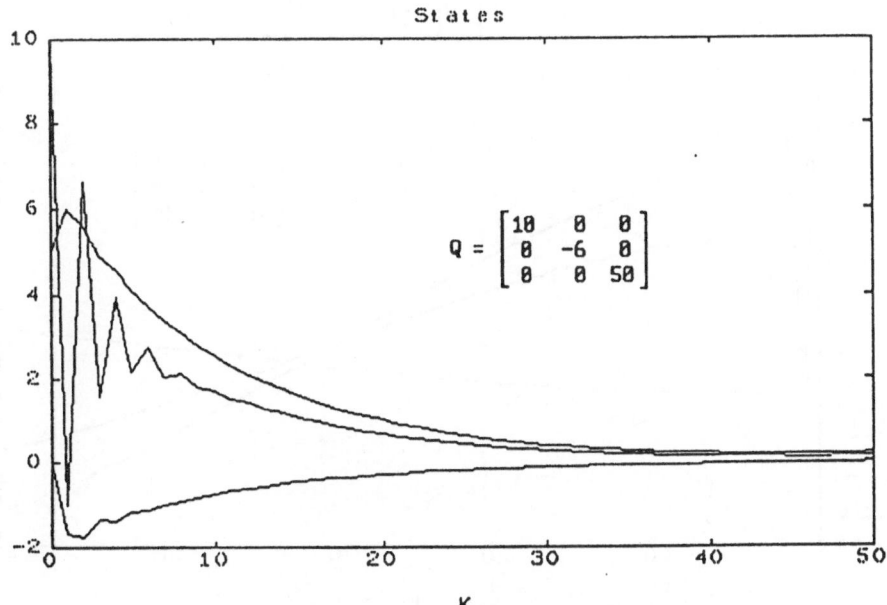

Figure 5.25a

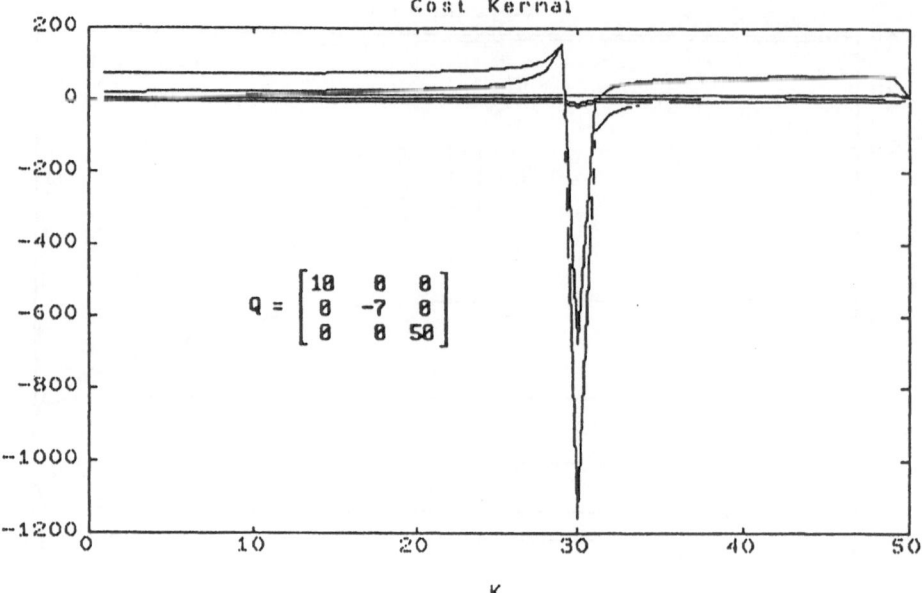

Figure 5.26a

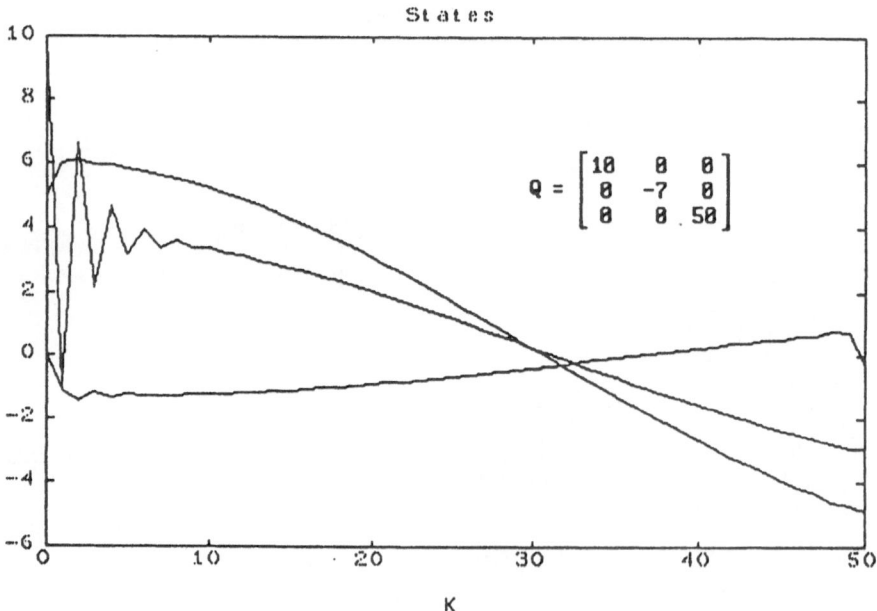

Figure 5.26b

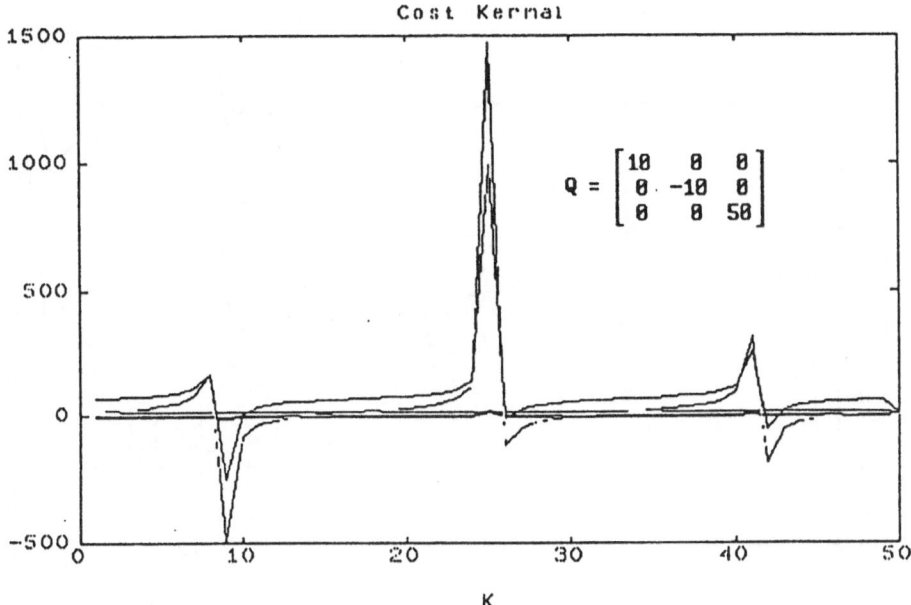

Figure 5.27a

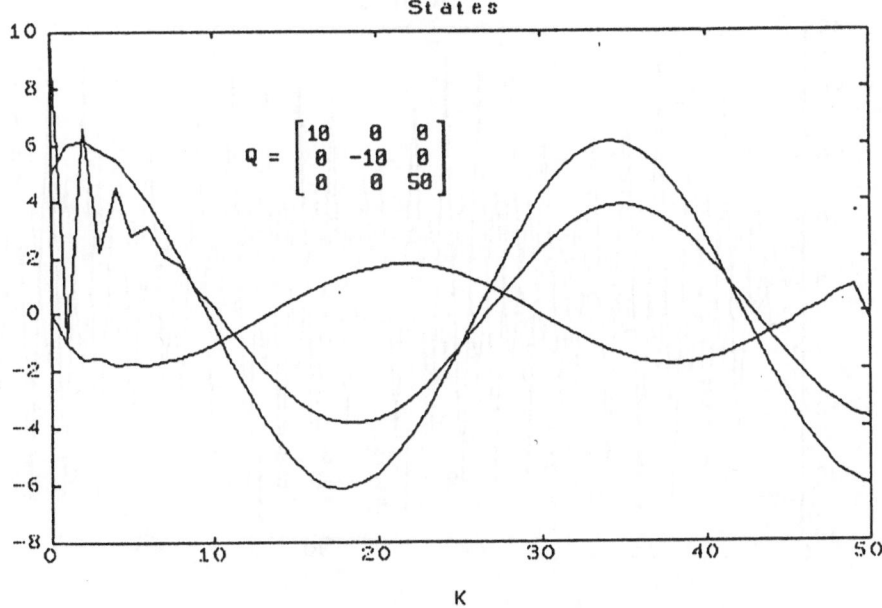

Figure 5.27b

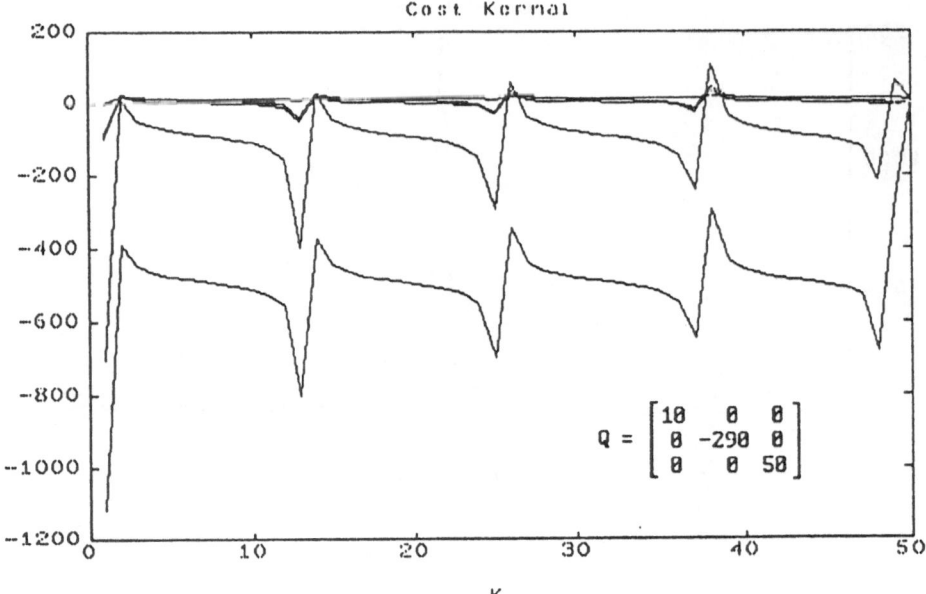

Figure 5.28a

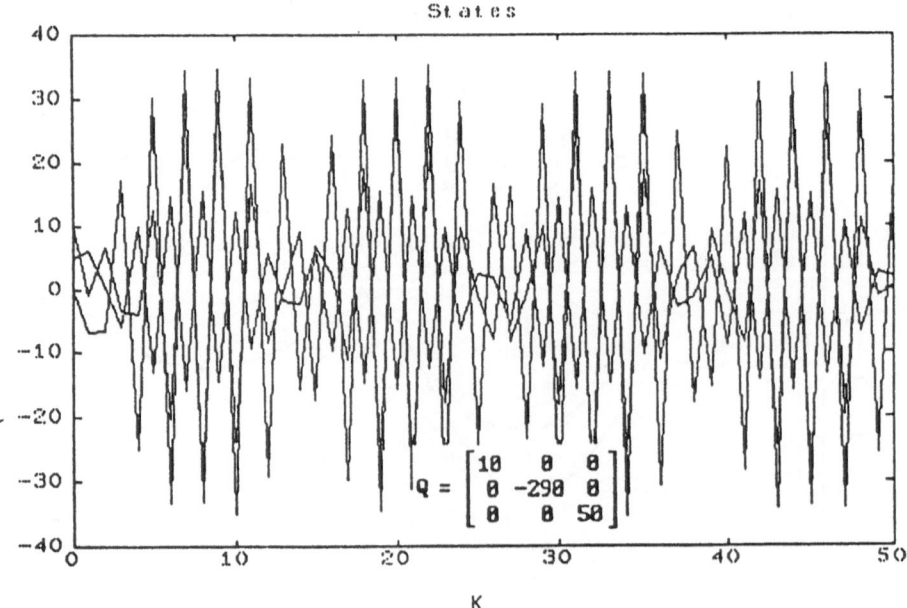

Figure 5.28a

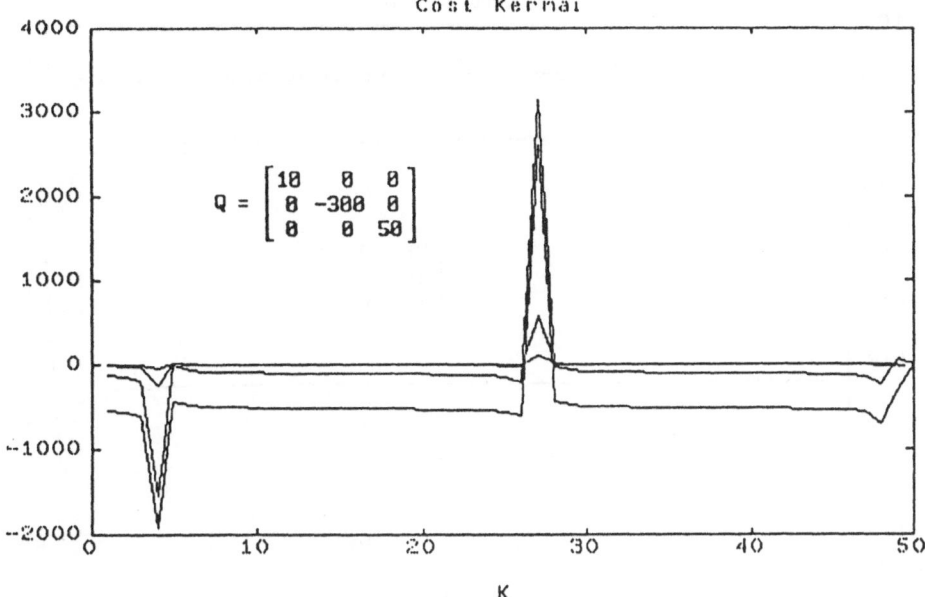

Figure 5.29a

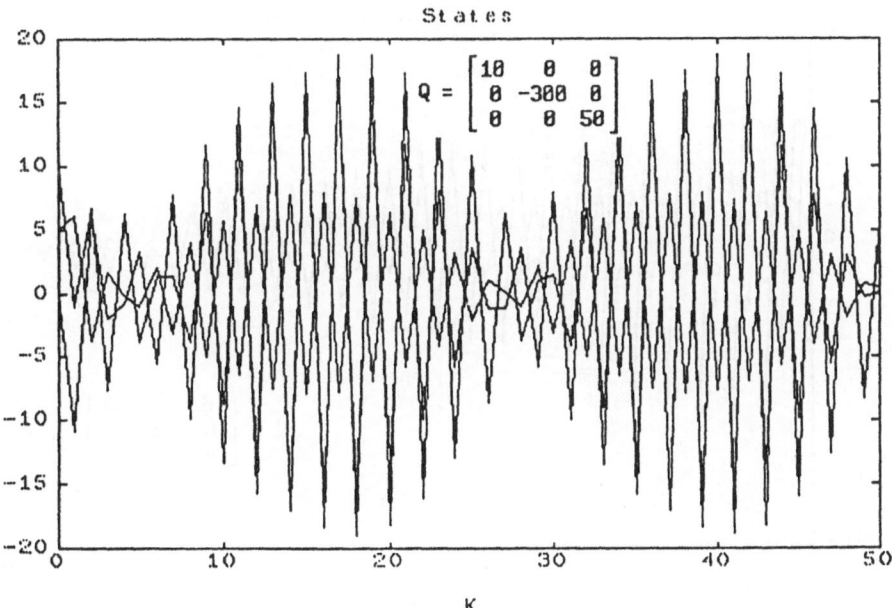

Figure 5.29b

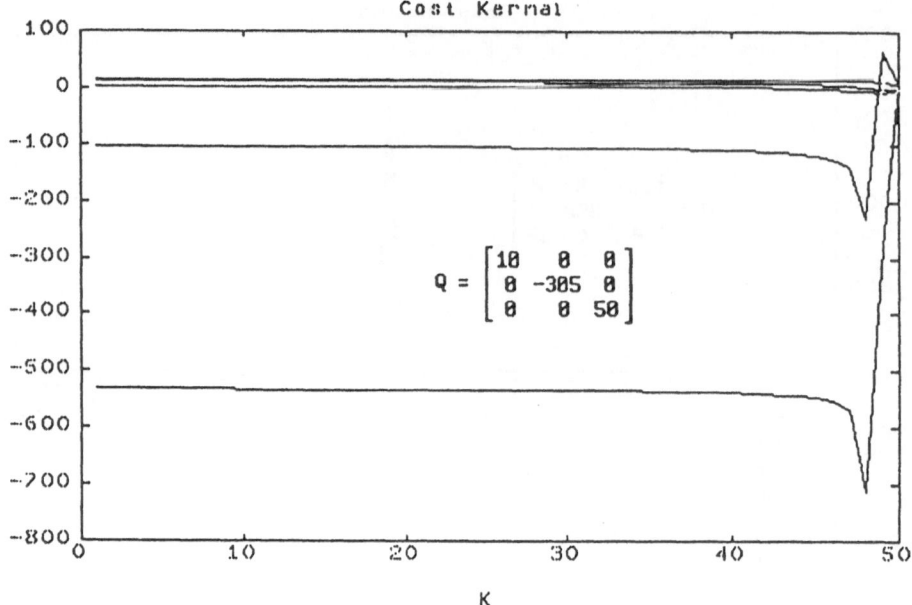

Figure 5.30a

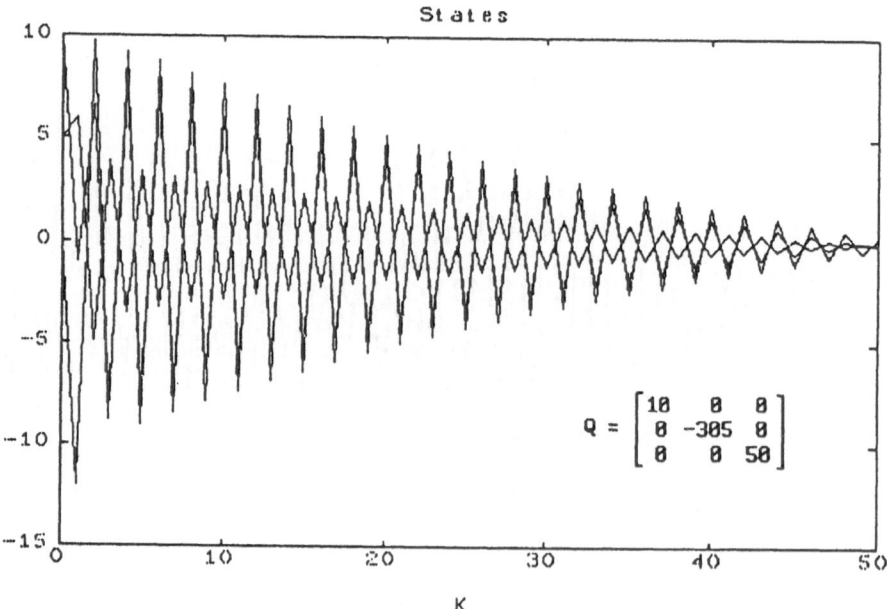

Figure 5.30b

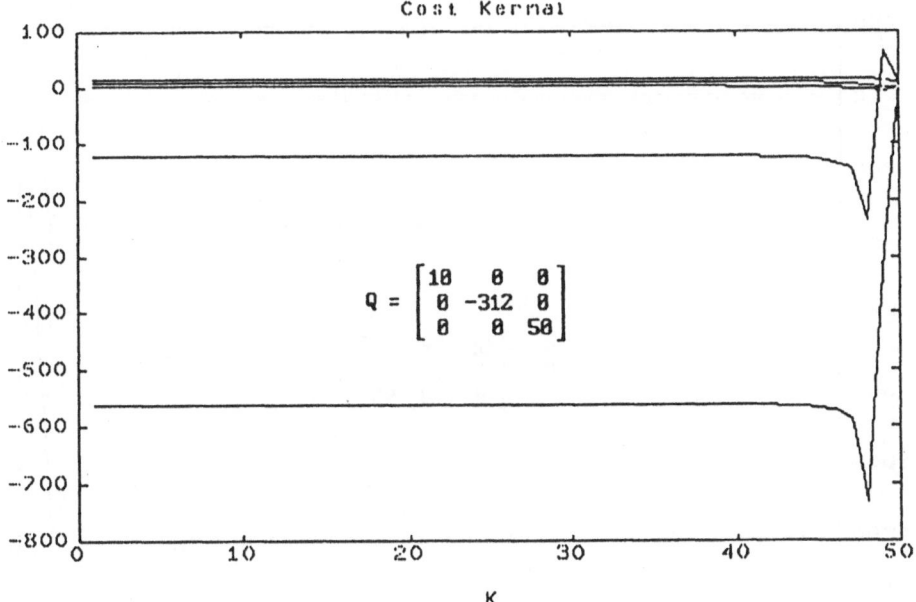

Figure 5.31a

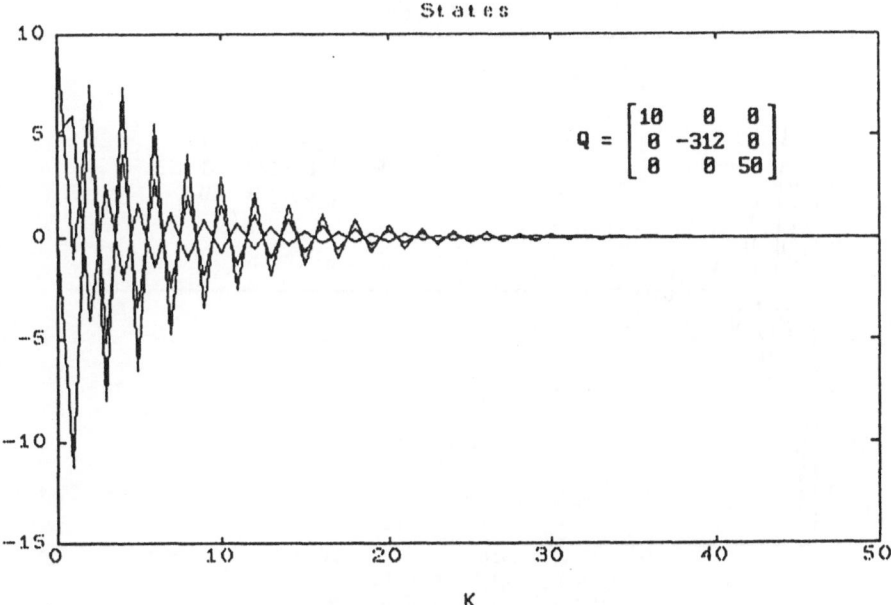

Figure 5.31b

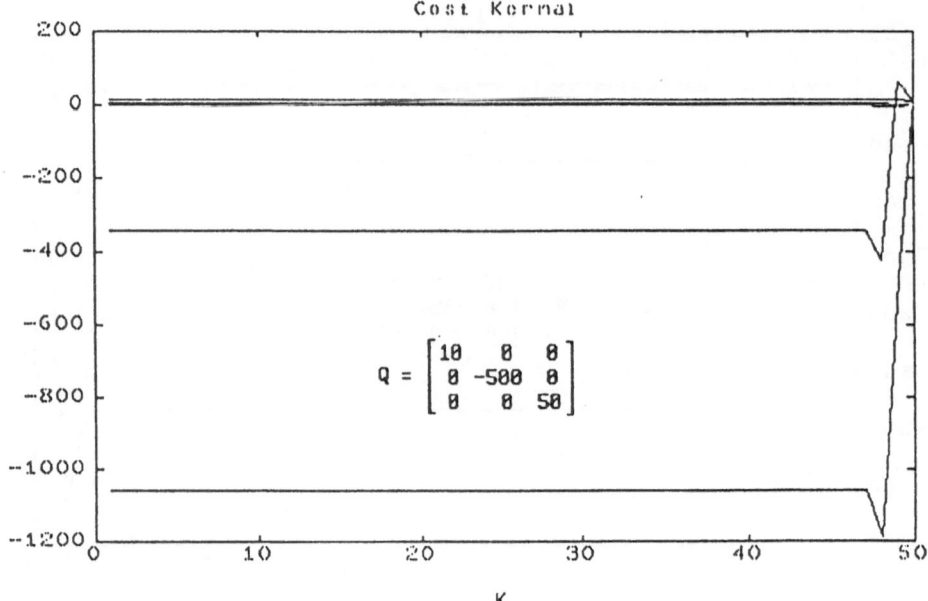

Figure 5.32a

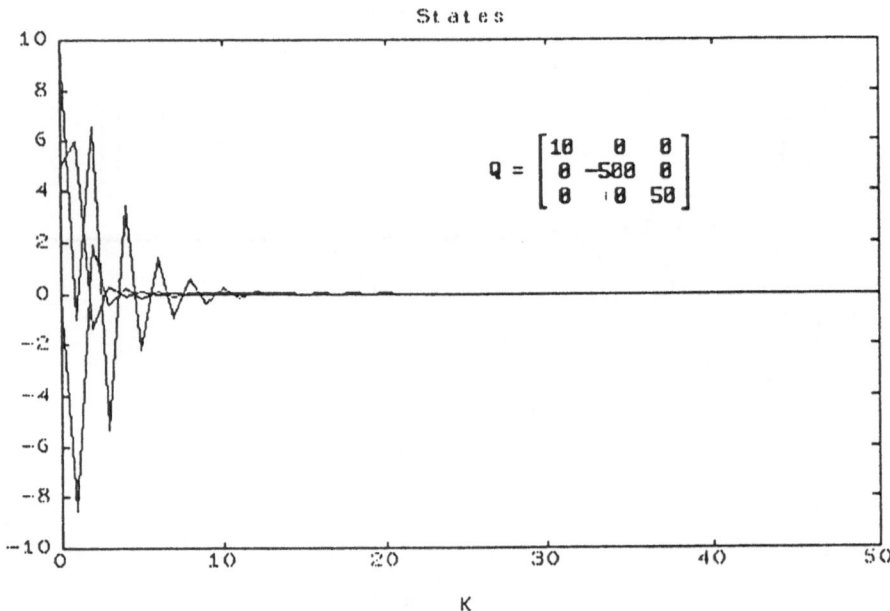

Figure 5.32b

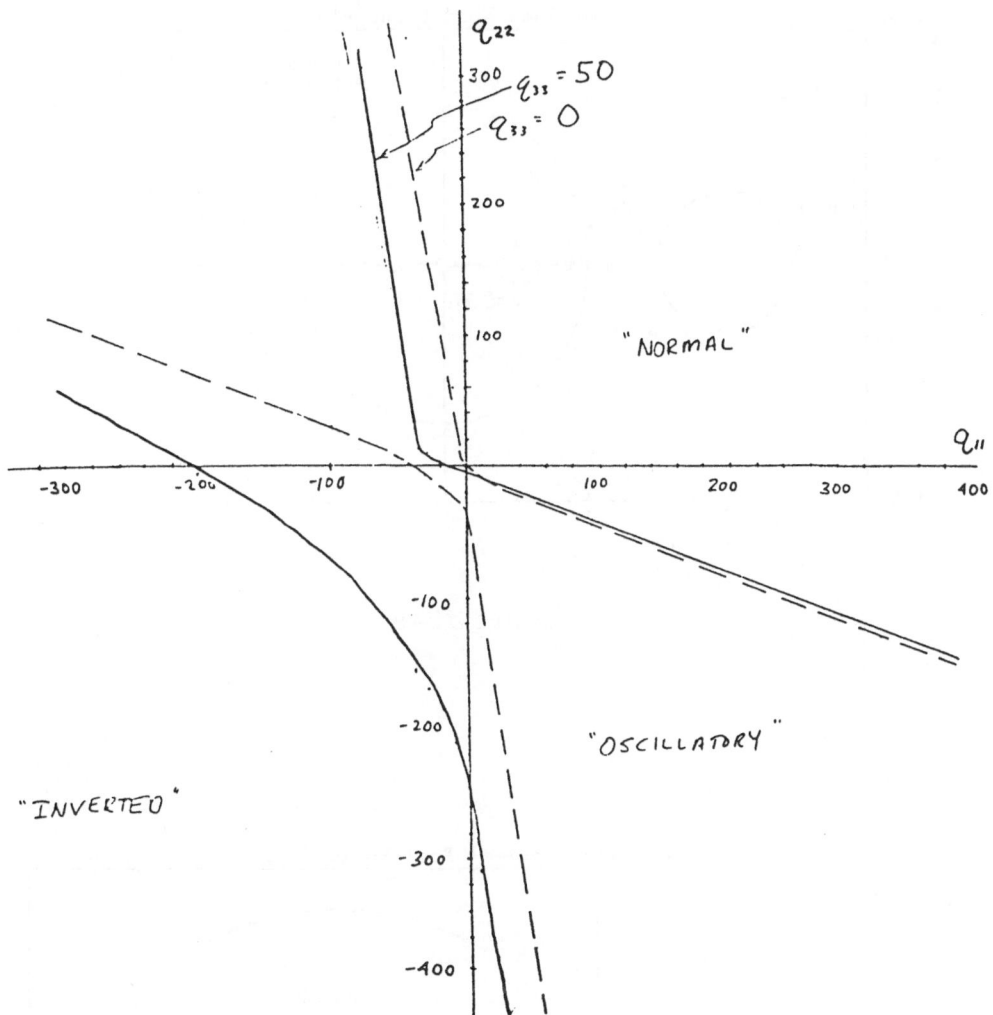

Figure 5.33 - Parameter region plot for the
3-dimensional system. Upper right is
normal operation. Lower left is inverted
operation. Middle regions are oscillatory.

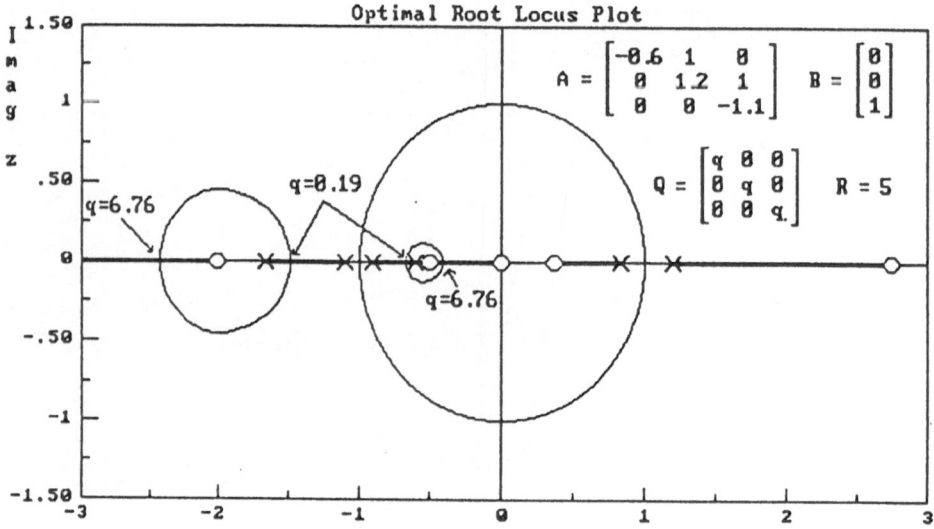

Figure 5.34a

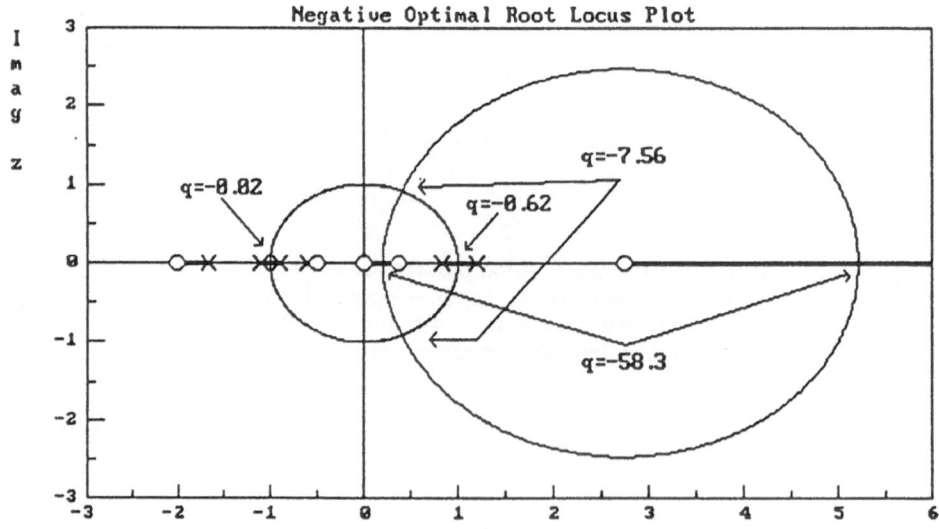

Figure 5.34b

CHAPTER SIX
CONCLUSION

This paper has contained a discussion of the hyperbolic iteration map and some of its applications. The application of most importance is the discrete linear quadratic regulator. It has been shown that solutions, meaningful or not, of the discrete linear regulator exist even for negative values of the state weighting matrix. This is demostrated most easily by the use of the Chang-Letov optimal root-locus technique. It is easily shown that for certain negative values of Q, the discrete optimal root locus will lie on, and move around, the unit circle. This will always be exemplified by "intermittent" behavior. This result is easily extended to continuous time systems, where the optimal root locus will always lie on the imaginary axis for some values of negative Q, again yielding "intermittent" behavior. Although a regulator design based on the negative values of Q does not appear to have any practical usage at this time, it is still interesting to observe that solutions still exist, and that their form is easily predictable using the Change-Letov method.

REFERENCES

Collet, P. and J. Eckmann (1981). *Iterated Maps on the Interval as Dynamical Systems*, Birkhauser, Boston.

Devaney, R.L. (1986). *An Introduction to Chaotic Dynamical Systems*, Benjamin/ Cummings, Menlo Park, CA.

Kirk, D.E. (1970). *Optimal Control Theory: An Introduction*, Prentice-Hall, Englewood Cliffs, NJ.

Lewis, F.L. (1986). *Optimal Control*, Wiley & Sons, New York.

May, R. (1976), "Simple mathematical models with very complicated dynamics," *Nature*, 10 June '76, pp. 459 - 467.

APPENDIX

INTRODUCTION TO CHAOS IN PHYSICAL SYSTEMS

INTRODUCTION

By his very nature the engineer tends to cringe at the word "chaos" and is rather frightened by its implications of "lack of order" or "lack of structure". However, a very new thrust of mathematical research, brought on by the widespread availability of high-powered computers, focuses on finding order in chaos and on understanding systems which have chaotic behavior. This research is of profound engineering importance; it is our goal here to answer, at a very simple level, the questions "What is chaos?" and "Why is the study of chaotic systems important?"

This goal is pursued as follows. The definition of chaos, which is "a sensitive dependence on initial conditions", is illustrated by studying the logistic equation, a simple discrete-time difference equation which is the most basic example of a system which can be chaotic. Qualitative examples, such as turbulent fluid flow, are then discussed in light of this new definition to further the reader's understanding. Following this, the mathematical requirements which any dynamical system must meet in order to exhibit chaotic behavior are presented. Finally, some motivations for the engineer's study of chaotic systems are put forth, and a specific example dealing with accuracy in digital simulation of analog system is pesented.

BACKGROUND: THE LOGISTIC EQUATION

For purposes of illustration we immediately restrict our attention to a first order discrete time system with its driver set equal to zero:

$$x_{k+1} = F\{ x_k, r_k \} \qquad \{A.1\}$$

$$x_k = \text{"state" of system}$$

$$r_k = \text{driver} = 0$$

Throughout most undergraduate engineering study, the function F in {A.1} is linear and time-invariant and there is always a neat, closed-form solution to {A.1} accompanied by various convenient "z-domain" analysis techniques. In such cases it can be shown that {A.1} will never exhibit chaotic behavior.

If F is nonlinear, however, then chaotic behavior can result, even in the simplest of systems. Equation {A.2}, known as the logistic equation, is such a system.

$$x_{k+1} = Ax_k(1 - x_k) \qquad \{A.2\}$$

"A" is a constant which falls within the range [0,4] (the system is unstable for "A" outside this range). The logistic equation {A.2} has several practical applications, one of its uses, for example, being a population growth model. The mapping x(k) into x(k+1) is shown in Figure A.1.

Since the system is nonlinear, there are no predetermined methods of analyzing its characteristics (e.g. finding its poles or zeroes) as there are for linear systems. To analyze

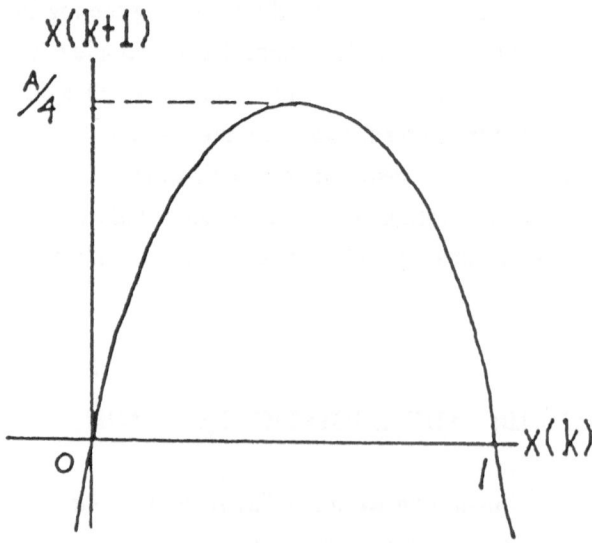

Figure A.1 - Plot of the logistic map.

this system, then, we must revert to simply finding "trends" in the system trajectory as we vary the parameter "A".

Setting the initial condition x(0) to a constant value, 0.6, let us examine the trajectories for some important values of A as we increase it from A = 0 to A = 4. Figure A.2, which is the system trajectory for A = 1.5, is a simple decay into a single constant steady-state value. This is also known as "period-1" behavior. Increasing A to 2.9, we observe a similar but more oscillatory decay to a constant steady-state value, as shown in Figure A.3. In general, this period-1 behavior exists for all values of A less than 3.

Further increasing A to A = 3.3 results in the period-2 steady-state trajectory of Figure A.4, and when A = 3.52 the trajectory is the period-4 steady state of Figure A.5. It seems that the behavior of the system is becoming more complicated as A increases.

To graphically represent the system in more condensed form the "density plot" for equation {A.2} is presented in Figure A.6. A density plot is simply the plot of all points hit by the steady-state trajectory of x(k) on the vertical axis versus the parameter "A" on the horizontal axis. For example, the period-2 trajectory of Figure A.4 shows up as the two-point "slice" corresponding to A = 3.3 in Figure A.6. Similarly, the period-4 case of Figure A.5 shows up as the four points above A = 3.52 in Figure A.6.

The chaotic regime of the logistic equation is very apparent upon examination of Figure A.6, especially the "weird" region in the parameter range A = 3.5 to A = 4.0, which is expanded and shown in Figure A.7.

Before defining the "chaos" itself, let us briefly discuss the "route to chaos" of the logistic equation as the parameter A increases. Starting at A = 1 to A = 3 the steady-state trajectory, the "attractor", is period-1 as previously noted. At A = 3 the system experiences a "bifurcation" to a period-2 attractor, and at A = 3.4 another such bifurcation occurs, resulting in a period-4 attractor. Such period-doubling behavior continues very rapidly as A increased, doubling to 8, 16, 32, 64, 128, ... until the point A = 3.57, when the period is 2^∞.

For values of A between 3.57 and 4.00, there is an infinite number of parameter windows $A = [A_{LOW}, A_{HIGH}]$ for which the attractor is periodic, and every integer period above 2 is represented in at least one interval. Examples of this are easily seen for periods 3, 5, and 6 in Figure A.7. In addition, there is also an infinite number of points $A_{c1}, A_{c2}, ...$ for which the trajectory of the system is truly "chaotic". Due to finite graphical precision in Figure A.7, however, it is difficult to differentiate between a "chaotic" point and a point which just has a high period.

What is the difference between a periodic trajectory and a chaotic trajectory? The answer to this question is the fundamental theme of chaos.

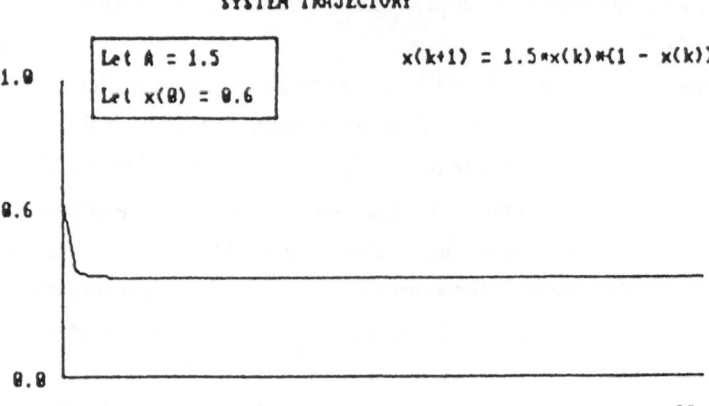

Figure A.2 - A typical logistic map trajectory, damped fixed point.

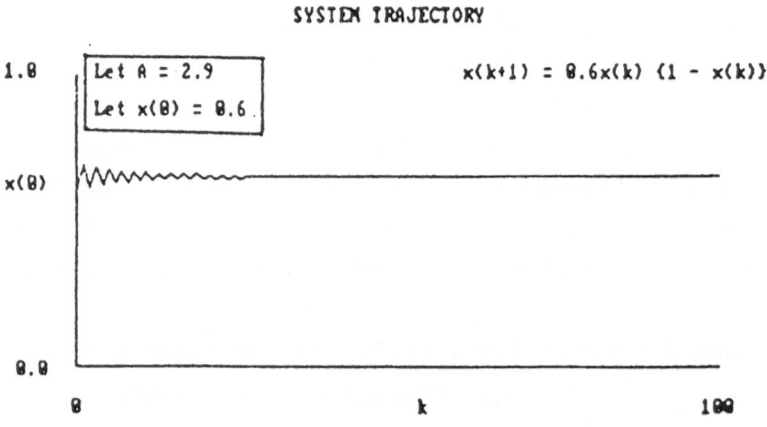

Figure A.3 - A typical logistic map trajectory, oscillatory fixed point.

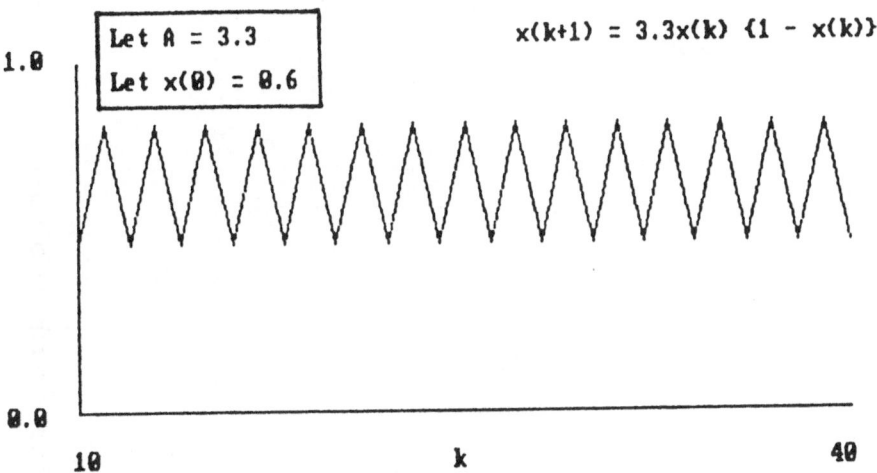

Figure A.4 - A typical logistic map trajectory, period 2 limit cycle.

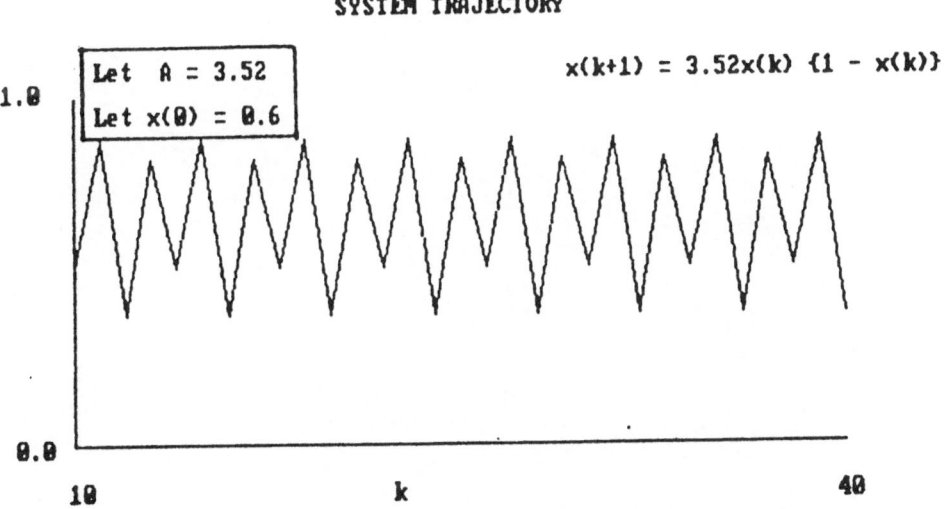

Figure A.5 - A typical logistic map trajectory, period 4 limit cycle.

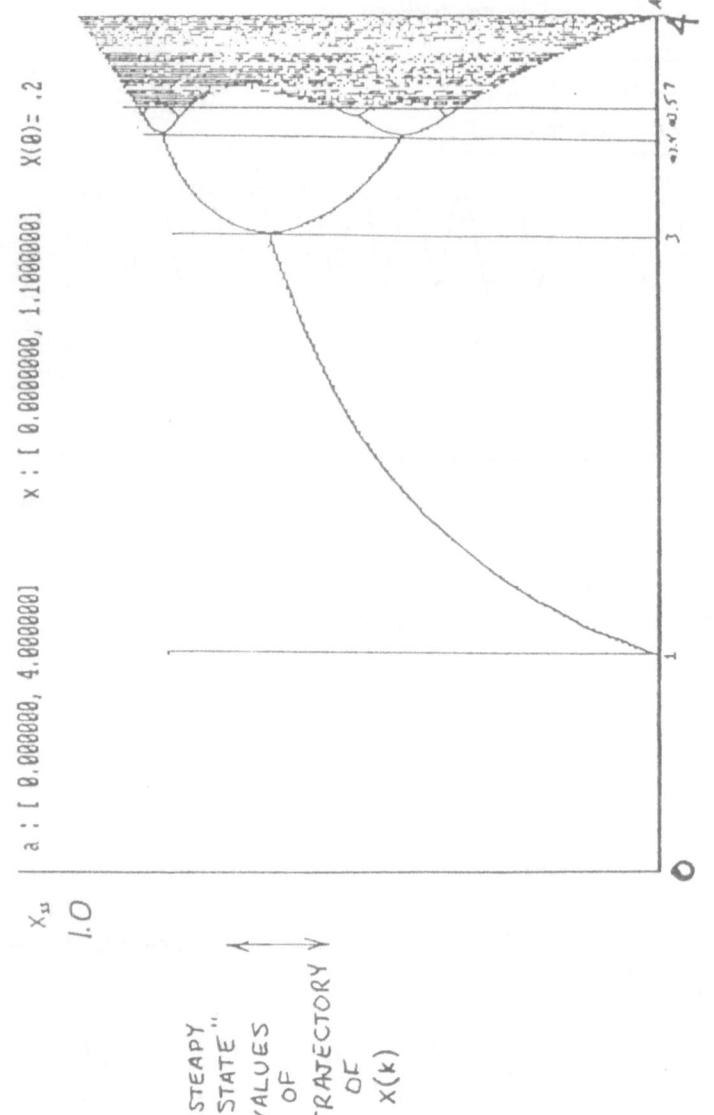

Figure A.6 - Bifurcation diagram for the logistic map.

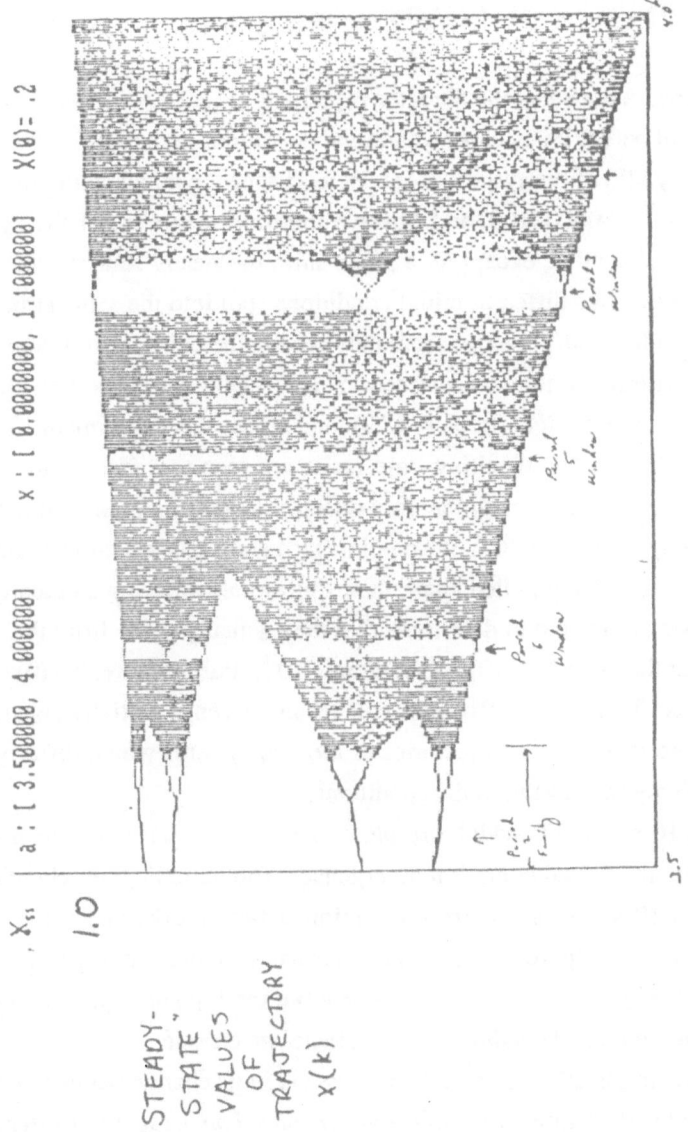

Figure A.7 - An expansion of Figure 6.

THE DEFINITION OF CHAOS

Let us choose A such that the attractor is periodic (choose A = 3.9066, which yields period 6). As we did before let us choose an initial condition, say x(0) = 0.600000000, and observe the trajectory x(k) of the system. Then let us change the initial conditionx(0) by only 10^{-8} so that x'(0) = 0.6000000100, and observe the trajectory x'(k) of the system. The results in Figure A.8 show that, except for a phase shift, the steady-state trajectories of x(k) and x'(k) are identical. Two different initial conditions "fall into the same attractor"; if we know the present condition x(0) we can accurately predict the behavior of the system for all time, even if there is error in the measurement of x(0). Now let us choose a value for A which is known to be chaotic (choose A = 4.00). and repeat this experiment. The result is shown in Figure A.9. The two resulting trajectories are seen to differ greatly and are not even close to each other, even though the only difference between them is that their initial conditions differed by a factor of 10^{-8}. Two points x(0) and x'(0), which were initially very close to each other, are quickly pulled apart, their trajectories differing so greatly that even after a short time we are unable to determine that they actually came from the same small region of space, that the quantity x(0) - x'(0) is only 10^{-8}. Furthermore, if x(0) - x'(0) were only 10^{-200}, outputs like Figure A.9 would still result. Even if initially separated by an *infinitesimal* nonzero amount, the trajectories of the two systems would differ greatly; the chaotic system is highly sensitive to initial conditions.

Figures A.10 and A.11, which are plots of x'(k) vs. x(k) with k as a parameter, graphically illustrate this initial condition divergence. The periodic, non-chaotic case A = 3.9066 in Figure A.10 shows a precise correlation between x'(k) and x(k): one can be predicted from the other. Figure A.11, on the other hand, shows that the trajectories x'(k) and x(k) are completely uncorrelated; when A = 4.00, the logistic equation {A.2} has a "sensitive dependence on initial conditions" and is therefore chaotic.

What are the implications of a system operating in its chaotic parameter regions? *It is impossible to predict the future output of the system based on present measurements*. This is because there is *always* measurement error and, even if this error is infinitesimal, the prediction based on iteration of its describing equation using this measurement will have *zero* correlation with the actual output of the system.

Although the system itself is deterministic (because it is exactly described by equation {A.2}), its output for future times based on present measurements is unpredictable and, therefore, is apparently random (within a given region of space). An equivalent definition of chaos, then, is "the apparent random behavior of a completely deterministic system".

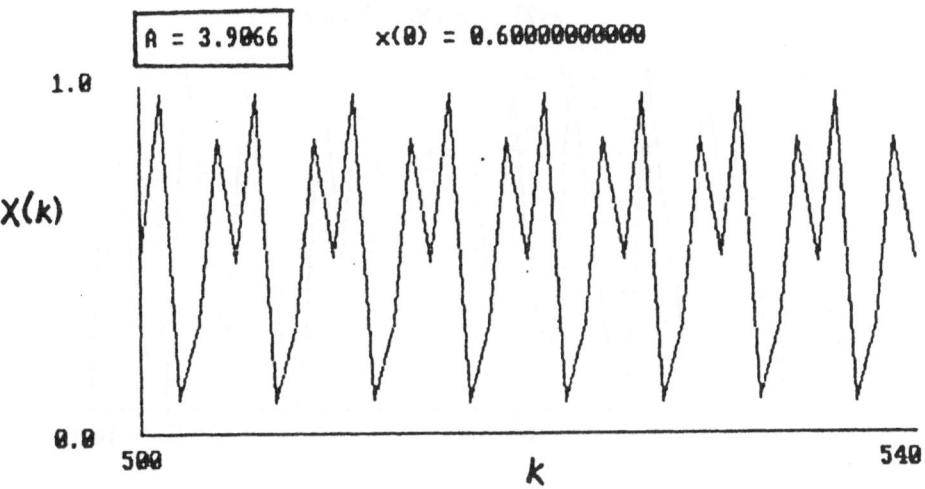

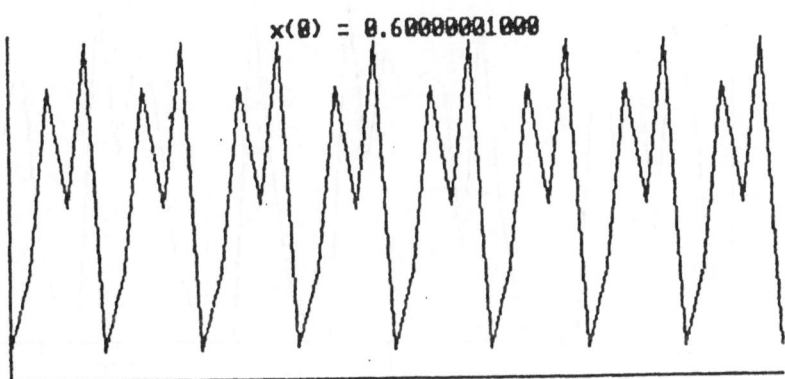

Figure A.8 - Logistic map, period 6; comparison of the two trajectories shows that the system is not sensitive to initial conditions.

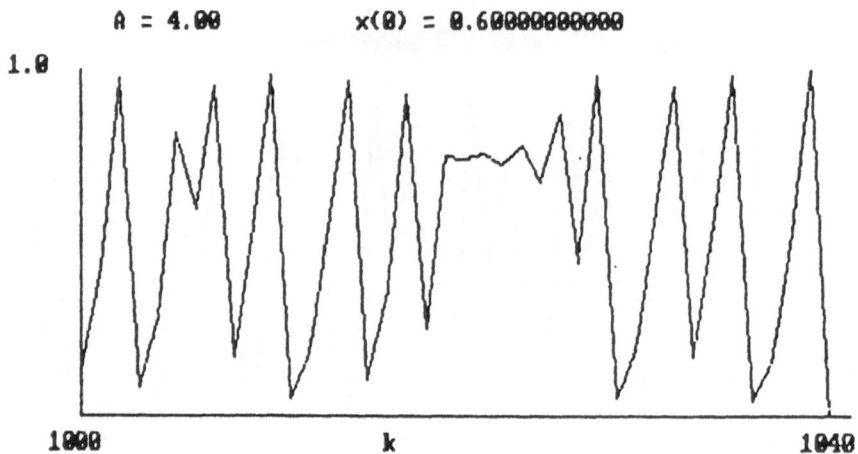

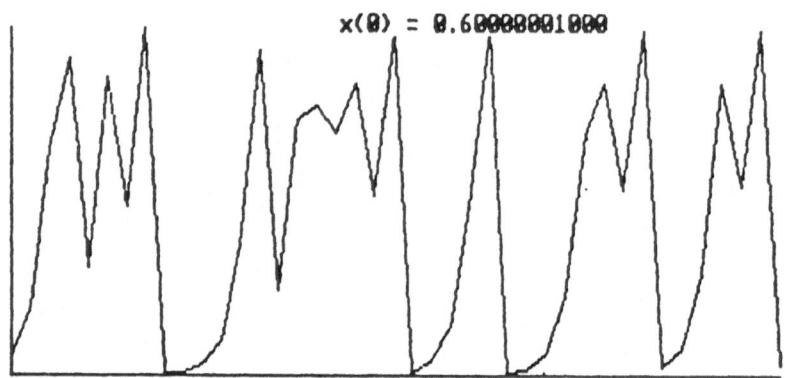

Figure A.9 - Logistic map, chaos; comparison of the two trajectories shows that the system is highly sensitive to any small change in initial conditions.

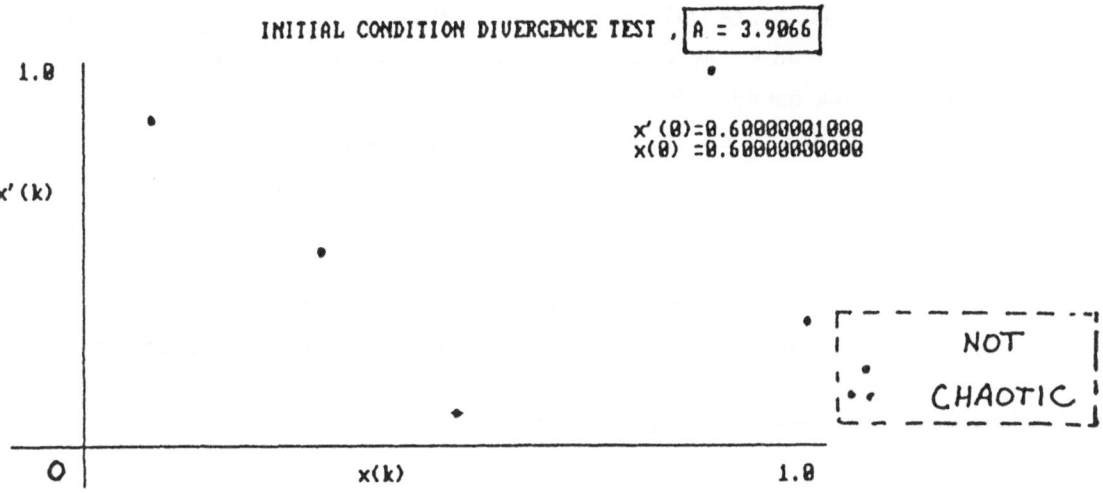

Figure A.10 - Initial conditions converge, no chaos.

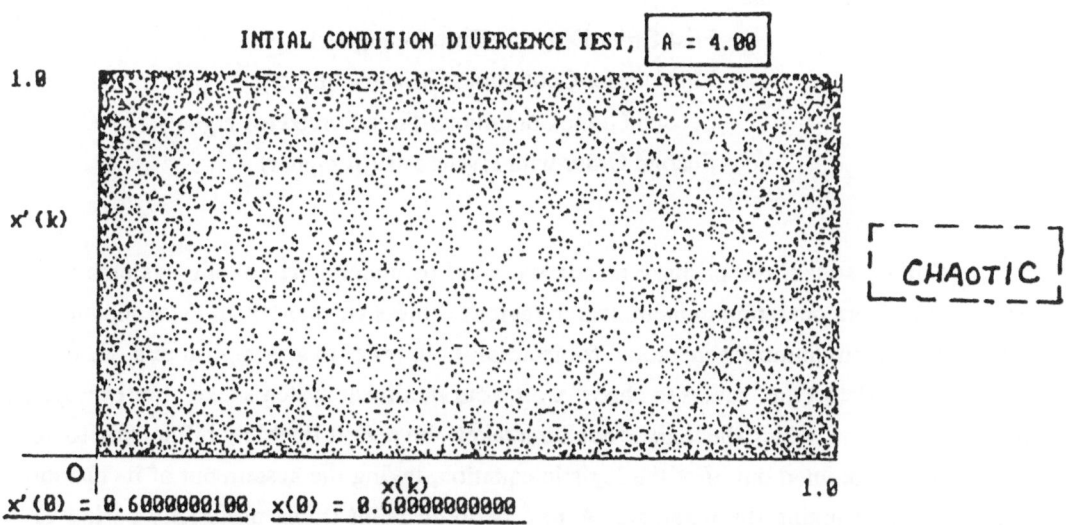

Figure A.11 - Initial conditions diverge, chaos.

These principles can be easily seen to tie in with our intuitive preconceptions of chaos. Consider the laminar flow of water (not chaotic) versus turbulent flow (chaotic). Under laminar flow, two initially adjacent water particles remain adjacent as they flow down the river; if their paths do diverge, they diverge predictably. In contrast, if the water flow is turbulent, their trajectories will be pulled apart so viciously that even a short distance down the river we are unable to determine that they came from the same initial region. Turbulent water flow is sensitive to initial conditions; it is chaotic.

The Earth's weather is another intuitive example of a chaotic dynamical system. The well-known "butterfly effect", which states that the mere flapping of a butterfly's wings could very well alter global weather conditions a short time later, is a prime example of sensitive dependence on initial conditions.

APPLICATIONS

In addition to being philosophically entertaining, awareness of chaos is important to the control systems engineer in a practical sense. As previously stated, chaos occurs in nonlinear systems, of which the real world is inherently composed. Put forth without proof, the following list gives several types of dynamical systems which can exhibit chaotic behavior:

Nonlinear systems:
-- First Order Discrete Time (or higher)
-- Third Order Continuous Time (or higher)
-- Second Order Continuous Time Forced (or higher)
-- Second Order with Hysteresis (or higher)

Simply stated, the primary reason we want to understand chaotic systems is to prevent them from operating in their chaotic regions (unless we want them to be chaotic, as in cases of random number generation, for example). In order to accomplish this, we must first be able to detect the "routes to chaos" which our nonlinear systems take. For example, the "route to chaos" which the logistic equation takes, the period-doubling bifurcation route, was previously pointed out. For the logistic equation, taking the system out of its chaotic region means changing the parameter A to a periodic point (most preferably period-1). Although much more complex than the simple logistic equation, analogous adjustments can be found for real-world chaotic systems to drive them out of their chaotic regions.

Let us turn to a concrete example to illustrate the above generality. It is desired to simulate on a digital computer the following nonlinear differential equation:

$$\dot{y} = By^2 + Cy \qquad \{A.3\}$$

From the conditions for chaos described previously, we know that {A.3} can never exhibit chaotic behavior, for it is only second-order continuous-time non-forced. Using Euler's method for digital simulation, however, yields

$$y_{k+1} = BTy_k^2 + (CT + 1)y_k \qquad \{A.4\}$$

(In this equation "T" is the step size.) Upon inspection of this equation it is clear that under a simple coordinate remapping it is nothing but the logistic equation, which is known to yield chaotic behavior! Thus, for certain parameters B, C, and T the simulation of {A.3} is chaotic, when we actually know that the system itself could never be: the simulation does not accurately reflect system behavior.

It is necessary, then, to be able to realize that the strange output of {A.4} is not due to a misbehaving system, but is actually due to a chaotic simulation of that system. No other explanation exists. Either the sampling interval must be decreased or, if this still does not bring {A.4} down to its period-1 region, another simulation method must be implemented.

Hence, in this example, an awareness of a nonlinear system's routes to chaos has lead to the recognition of an erroneous digital simulation of an analog system, and has given insight into correcting the error. Again, although this was a very simple example, it is indicative of what can happen with more complicated, higher-order nonlinear systems.

CONCLUSIONS

In brief, this appendix has exposed the reader to the principle of chaos through the study of a simple dynamical system which exhibits chaotic behavior. In addition, reasons have been provided as to why the study of such systems is of engineering importance. The principal conclusions are as follows:

(1) When applied to control systems, the term "chaos" signifies a "sensitive dependence on initial conditions," or, equivalently, "the apparent random behavior of a completely deterministic system".

(2) Chaos abounds in the real world; generally speaking, any multidimensional nonlinear system may exhibit chaotic behavior.

(3) The understanding of the routes to chaos taken by nonlinear systems, or at least an awareness of the existence of chaos in such systems, is of vital importance in their accurate control and simulation. In the latter case, an improper simulation can falsely indicate the misbehavior of a working system due to the chaotic nature of the simulation itself.

Lecture Notes in Control and Information Sciences

Edited by M. Thoma and A. Wyner

Lecture Notes in Control and Information Sciences

Edited by M. Thoma and A. Wyner

Lecture Notes in Control and Information Sciences

Edited by M. Thoma and A. Wyner